Proceedings of the 14th Sino-Russia Symposium on Advanced Materials and Technologies (Abstracts)

第十四届

中俄双边新材料新工艺研讨会论文集（摘要）

中国有色金属学会　编

By the Nonferrous Metals Society of China

北　京

冶 金 工 业 出 版 社

2017

内 容 提 要

本书收录了第十四届中俄双边新材料新工艺研讨会提交的全部论文，内容涵盖了纳米材料、稀有金属和合金、新能源和储能材料、电子信息材料、生物医用材料、航空航天材料、表面工程技术、陶瓷复合材料、新型冶金工艺和材料加工技术等领域。

本书可供材料及其加工领域的科研、技术和管理人员阅读，也可供大专院校有关师生参考。

Brief Introduction

The book contains all the papers submitted to the 14th Sino-Russia symposium on advanced materials and technologies. The topic to be discussed covers from nanomaterials, rare metals and alloys, new energy and energy-saving materials, electronic information materials, biological and polymer materials, aerospace materials and surface engineering technology, ceramic and composite materials, new metallurgical processes to materials processing technology.

This book can be used for researchers, technicists and managers in materials and processing fields and teachers and students in universities.

图书在版编目(CIP)数据

第十四届中俄双边新材料新工艺研讨会论文集：摘要 = Proceedings of the 14th Sino-Russia Symposium on Advanced Materials and Technologies (Abstracts) / 中国有色金属学会编. —北京：冶金工业出版社, 2017.11

ISBN 978-7-5024-7654-0

Ⅰ.①第… Ⅱ.①中… Ⅲ.①新材料应用–国际学术会议–文集 Ⅳ.①TB3-53

中国版本图书馆 CIP 数据核字 (2017) 第 255342 号

出 版 人 谭学余

地　　址 北京市东城区嵩祝院北巷 39 号 邮编 100009 电话 (010)64027926

网　　址 www.cnmip.com.cn 电子信箱 yjcbs@cnmip.com.cn

责任编辑 杨盈园 王梦梦 美术编辑 杨 帆 版式设计 孙跃红

责任校对 王永欣 责任印制 牛晓波

ISBN 978-7-5024-7654-0

冶金工业出版社出版发行；各地新华书店经销；北京建宏印刷有限公司印刷

2017 年 11 月第 1 版, 2017 年 11 月第 1 次印刷

210mm×297mm；5.75 印张；180 千字；72 页

169.00 元

冶金工业出版社 投稿电话 (010)64027932 投稿信箱 tougao@cnmip.com.cn

冶金工业出版社营销中心 电话 (010)64044283 传真 (010)64027893

冶金书店 地址 北京市东四西大街 46 号(100010) 电话 (010)65289081（兼传真）

冶金工业出版社天猫旗舰店 yjgycbs.tmall.com

（本书如有印装质量问题，本社营销中心负责退换）

编 委 会

Editorial Board

前　言

中俄双边新材料新工艺研讨会是由中国有色金属学会与前苏联科学院巴依科夫研究院、强度物理及材料科学研究所于 1991 年共同倡议组织的每两年轮流在中国和俄罗斯召开的国际学术会议，至今已召开了十三届，会议的规模和范围逐步扩大。研讨会的召开有力地促进了中俄两国科技人员的技术交流和合作，已成为两国冶金材料界具有代表性的学术会议。本届研讨会得到了两国政府和中俄两国共同组建的“二十一世纪和平、友好、发展委员会”的支持，将于 2017 年 11 月 28 日~12 月 1 日在中国海南省三亚市举办。

会议旨在为研究人员提供一个自由交流的平台，共同分享新的思路、新的创新和解决方案。主题是“金属、陶瓷与复合材料”，内容涵盖了纳米材料、稀有金属和合金、新能源和储能材料、电子信息材料、生物医用材料、航空航天材料、表面工程技术、陶瓷复合材料、新型冶金工艺和材料加工技术等领域。此外，会议组织者将邀请知名专家、学者作大会专题报告，所有参会者将有机会与报告人面对面讨论，这对参会者进行更深入的科学讨论和增进相互了解非常有帮助。

此次会议得到了相关单位和人员的大力支持。在此，我们表示衷心感谢；其次，要感谢论文作者们的积极参与；再次，也要对所有的组织委员会成员和审稿人等付出的辛勤工作表示感谢。希望所有的参会者都能提出宝贵意见和建议，以便提高工作效率和服务。预祝此次会议圆满成功！

陈金训

大会主席

国务院参事、中国有色金属工业协会会长

2017 年 10 月

Preface

The international conference, Sino-Russia International Symposium on New Materials and Technologies, is organized by *Nonferrous Metals Society of China*, *Baikov Institute of Metallurgy and Materials* and *Institute of Strength Physics and Materials* of the former Soviet Union starting in 1991. It was held alternately in China and Russia every two years, it has been successfully held thirteen times with gradually expanded scale and scope. Moreover, it effectively promoted the development of scientific and technical Sino-Russian cooperation and has become a representative conference in metallurgical materials industry between the two countries. The SRISNMT 2017 was supported by the two governments and the "21st Century Peace, Friendship and Development Committee". It will be held from November 28 to December 1, 2017 in Sanya, Hainan Province, China.

The goal of the conference is to provide researchers with a free exchange forum to share the new ideas, new innovation and solutions with each other. The topic to be discussed cover abroad field ranging from nanomaterials, rare metals and alloys, new energy and energy-saving materials, electronic information materials, biological and polymer materials, aerospace materials and surface engineering technology, ceramic and composite materials, new metallurgical processes to materials processing technology.In addition, the conference organizers will invite some famous keynote speaker to deliver their speech in the conference. All participants will have the chance to discuss with the speakers face to face, which is very helpful for participants to deeper scientific discussions and promote mutual understanding.

During the planned stage of the conference, we have got help from different people, different departments, and different institutions. Here, we would like to express our first sincere appreciation for their kind and enthusiastic help and support for our conference. Secondly, the authors should be thanked too for their enthusiastic writing attitudes towards their papers. Thirdly, all members of program secretariat, reviewers and program committees should also be appreciated for their hard work.

In a word, it is the different teams efforts that will make our conference be successful on November 28 to December 1, 2017 in Sanya. We hope that all of participants can give us good suggestions to improve our working efficiency and service in the future, and we also hope to get your supporting all the way.

Quanxun Chen
General Chair
State Department Counselor, The President of China Nonferrous Metals Industry Association
October, 2017

Contents

The Studies on the Interfacial Evolution Behavior of $ZrO_2(ZrB_2)$ Active Diffusion Barrier
······ Lintao Liu, Zhengxian Li, Yunfen Chen 1

Mathematical Modeling and Computer Simulation of Steelmaking Technologies
······ Grigorovich K.V., Komolova O.A., Gorkusha D.V. 1

Development of Molybdenum-based Metallic Glasses
······ Gao Xuanqiao, Li Bin, Zhang Wen, Ren Guangpeng, Xue Jianrong, Li Laiping 2

Investigations of Decarburisation Rates of High Chromium Iron Base Melts by Low-temperature Plasma. Mathematical Model and Experiments ······ K. V. Grigorovich, O. A. Komolova, B. A. Rumyantsev 2

Advances of Mo-Re Alloys on Microstructures and Mechanical Properties
······ Li Laiping, Gao Xuanqiao, Hu Zhongwu, Liang Jing, Lin Xiaohui, Xue Jianrong 3

Development of Methods for Preparation of Oxide Ceramic Systems via Individual Precursors
······ V.M. Novotortsev, Zh.V. Dobrokhotova, A.V. Gavrikov, P.S. Koroteev, N.N. Efimov 3

Effect of Multilayered Structure on the Properties of Ti/TiN Coating
······ Yanfeng Wang, Zhengxian Li, Jihong Du, Haonan Wang, Changwei Zhang 4

Synthesis and Investigation of the $Cd_{1-x}Fe_xCr_2S_4$ Solid Solutions as a Prospective Basis for the Magnetoelectrical and Magnetocapacitive Materials
······ V.M. Novotortsev, T.G. Aminov, G.G. Shabunina, N.N. Efimov 4

Current Research in Silver Nanowires ······ Yongqian Wu, Kunkun Chen, Bosheng Zhang, Qigao Cao 5

Comparative Kinetics of Extraction of Molybden-containing Components from the Waste Hydrotreatment Catalyst by Mineral Acids ······ S.P. Perekhoda, E.Yu. Nevskaya, O.A. Egorova 5

Fabrication of Ag Sheathed FeSe Superconducting Tapes
······ Shengnan Zhang, Chengshan Li, Jianqing Feng, Jixing Liu, Botao Shao, Pingxiang Zhang 6

Synthesis and Optical Properties of Aluminum Oxynitride (Alon) Doped with Rare-earth Metals Ions EU^{2+} and CE^{3+}
······ N.S. Akhmadullina, A.S. Lysenkov, V.V. Yagodin, A.V. Ishchenko, Yu.F. Kargin, B.V. Shul'gin 6

Preparation and Mechanical Properties of SiC Whisker Reinforced Multi-phase Mo-Si-B Alloy

…… Bin Li, Laiping Li, Xiaohui Lin, Huiqing Fan, Pingxiang Zhang　7

Magnetic Properties of the Hard Magnetic Alloy of FE-27CR-10CO-2MO …… Alymov M. I., Milyaev I.M.,Yusupov V. S., Zelensky V.A., Milyaev A.I., Ankudinov A.B., Abashev D. M.　7

Effect of Strain Rate on Texture Evolution of TLM Titanium Alloy During Hot Deformation …… X.F. Bai, Y.Q. Zhao, X.M. Wang, J.Y. Liu　8

Microstructure of the Polycrystalline Alloys of the $Al_{85}Ni_{11-x}Fe_xLa_4$ System …… N.D. Bakhteeva, N.N.Kolobylina, E.V.Todorova, A.G.Ivanova, A.L. Vasiliev　8

Preparation and Properties of Foam Titanium for Biological Implants …… Li Zengfeng, Tang Huiping, Chen Gang, Tan Ping, Zhao Shaoyang, Shen Lei　9

Fiber-optical Acoustic Emission Sensors for Registering Damage in Polymeric Composite Materials …… Bashkov O.V., Romashko R.V., Zaikov V.I., Khon H., Bashkov I.O., Bryansky A.A.　9

Effects of Heattreatments on the Properties of Coarse-grained WC-10 Cocemented Carbides with Low Carbon Content …… X.C. Xie, Z. W. Li, R. J. Cao, Z. K. Lin　10

Acoustic Emission of Fatigue Damage in Structured Pure Titanium VT1-0 …… Bashkov O.V., Bashkova T.I., Popkova A.A., Sharkeev Yu.P., Eroshenko A.Yu., Tolmachev A.I.　10

Film-forming Characteristic of Lead-silver-calcium Alloy Anode during Zinc Electrowinning …… H.T. Yang, C.L. Fan, Y.G. Wei, B.M. Chen, Z.C. Guo　11

Hydrogen Effect on the Martensitic Transformations and the Superelasticity of TiNi Based Alloy with Ultrafine-grained Structure …… Baturin A.A., Lotkov A.I., Grishkov V. N., Rodionov I.S.　11

Selective Hydrolysis and Removal of Trace Titanium during the Chlorination Process for High Purity Vanadium Pentoxide …… Chuanlin Fan, Haitao Yang, Qingshan Zhu　12

Reference Complex for Metrological Ensuring of Ultrasonic Measurements in Solid Media …… P.V. Bazylev, I.Y. Krumgolz, V.A. Lugovoy　12

Study on Crack Propagation Rate of 600MPa High Strength Steel for Mn-Ti and Nb-V System …… Li Xiaolin, Cui Yang, Yang Xiaohe, Jin Zhao　13

Spark Plasma Sintering of High-strength Fine-grained Alumina Ceramics …… Boldin M.S., Chuvil'deev V.N., Popov A.A., Nokhrin A.V.　13

Effect of GH4169 Alloy Powder Characteristics on Selective Laser Melting Process …… DU Kaiping, SHEN Jie, ZHANG Shuting, YU Yueguang　14

Investigation of the Processes Induced by High-dose Ar^+ Implantation into Vanadium and Vanadium-based Alloy V-4.51 Ga-5.66 Cr …… I. V. Borovitskaya, S. N. Korshunov, A. N. Mansurova, N. A. Vinogradova, A. V. Mihajlova, V. V. Paramonova　14

The Experimental Investigation on the Flow Property of Aluminum Alloy AA6016 with the Fluid Media
······ Lihui Lang, Quanda Zhang, Sergei Alexandrov 15

The Molybdenum Alloys for Welded Vacuum-tight Glass-to-metal Parts of Electronic Vacuum Devices
······ Burkhanov G. S., Kirillova V. M., Minina N. A., Kadyrbayev A. R., Mikhailov B. P., Sdobyrev V. V., Dement'ev V. A., Dormidontov N.A., Korenovskiy N. L., Sevostyanov M. A., Baikin A. S. 15

Elastic Mechanisms for a C17200 Alloy
······ PENG Lijun, HUANG Guojie, XIE Haofeng, FENG Xue, YANG Zhen 16

Design and Investigation of III-V Solar Cells on Silicon Substrate with YSZ Anti-reflective Coating
······ Alexander Buzynin, Yuri Buzynin, Vladimir Shengurov, Nickolai Baidus, Nuofu Chen 16

Effect of Heat Treatment on the Bending Behavior of Cu/Al Composite Wires
······ Zhen Yang, Haofeng Xie, Xujun Mi, Guojie Huang, Lijun Peng, Xue Feng, Xiangqian Yin 17

Evolution of the Phase Morphology of Disperse Systems Consisting of Particles of Clay Minerals under the Action of MW Radiation
······ Chetverikova A.G., Filyak M.M., Kanygina O.N., Berdinsky V.L., Savinkova E.S. 17

Discussion on Deformation of High Strength Aluminum Alloy by Mesoscopic Mechanics
······ ZHU Qifang, SUN Zeming, Fan Zhigang, Li Pu, Xia Wen, Liu Shufeng, SHARKEEV Yurii, Anna Eroshenko, KOMAROVA Ekaterina 18

The Parallel Algorithm of Multi-billion Atom Samples Generation for Molecular Dynamics Simulations
······ A.M. Chuprunov, F.A. Sapozhnikov 18

Preparation of Nanometer Yttrium Oxide by Precipitation Method
······ Li Xingying, Wang Lei, Liu Zhiqiang, Jin Mingya 19

Deoxidation Equilibrium in Iron-nickel Melts
······ V.Ya. Dashevskii, A.A. Aleksandrov, A.G. Kanevskii, L.I. Leont'ev 19

Comparison of Ag and Zr with Same Atomic Ratio in Cu-Cr Alloy
······ Feng Xue, Xie Haofeng, Li Zongwu, Mi Xujun, Huang Guojie, Peng Lijun, Yang Zhen, Yin Xiangqian 20

"Vikhr" and Another DPF Devices for Fuson Application Experiment ······ E.V. Demina, V.A. Gribkov, A.S. Demin, S.A. Maslyaev, E.V. Morozov, M. Paduch, V.N. Pimenov, M.D. Prusakova, E. Zielinska 20

Green Evaluation of Microwave Heating Furnace Using Fuzzy Decision-making Method and Life Cycle Assessment Methods ······ Xiaoying Zheng, Jin Chen, Guo Chen, Jinhui Peng, Rongsheng Ruan 21

The Problems of Obtaining Multilayer Nanowires Ni/Cu
······ I.M.Doludenko, D.L. Zagorskiy, S.A. Bedin, G.G.Bondarenko, V.V.Artemov 21

Preparation and Characterization of Single Crystal Type $LiNi_{0.8}Co_{0.1}Mn_{0.1}O_2$ as Cathode Material for

Lithium-ion Batteries ········ Xuequan Zhang, Minchao Ru, Yanbin Chen 22

Influence of the Millisecond Laser Irradiation on the Morphology and Crystallization of a Thermo-activation Zone in Bulk Amorphous Alloy Zr-Cu-Ag-Al ········ Fedorov V.A., Shlykova A.A., Gasanov M.F. 22

Effect of Fe_2O_3 Content on Thermal Insulation Properties of YSZ Thermal Barrier Coatings ········ Ji Xiaojuan, Yu Yueguang, Wei Yudao, Li Yujie, Gao Lihua 23

Biologically Active Agents Based on Colloidal Selenium ········ Folmanis G.E., Kovalenko L.V., Fedotov M.A., Volchenkova V.A. 23

Microstructure and Residual Stress Distribution of TA15/BTi6431S Titanium Alloys Welding Joints by Ultrasonic Impact Treatment ········ Xiaoyun Song, Guanglu Qian, Wenjun Ye, Songxiao Hui 24

Impulse Activation of Hydrogen-air Fuel Cell ········ Elena Galitskaya, Ekaterina Gerasimova, Yuri Dobrovolsky, Alexander Sivak, Vitaly Sinitsyn 24

Quasi-static and Dynamic Properties of Ti-4Al-3V-0.6Fe-0.2O Titanium Alloy Plates ········ Rui Liu, Songxiao Hui, Wenjun Ye, Yang Yu, Xiaoyun Song 25

Layer Laser Fusing of Sm and Fe Powders on the Non-magnetic Stainless Steel Substrates with and Without Additional Magnetic Field: Composition and Ferromagnetic Properties ···· N.G. Galkin, Y.N. Kulchin, E.P. Subbotin, A.I. Nikitin, D. S. Yatsko, M.E. Stebliy, A.V. Nepomnyaschiy 26

The Development Trend of Zinc Electrodeposition Technology ········ Xie Gang, Hu Fengying, Ma Jilun 27

Investigations of Young's Modulus and Microhardness of Laser-fused Sm-Fe and Sm-Co Coatings on Two Types of Substrates ········ N.G. Galkin, Y.N. Kulchin, D. S. Yatsko, E.P. Subbotin, A.I. Nikitin, A.V. Nepomnyaschiy 27

Investigation on High Performance (γ+α_2+B2) Multi-phase TiAl Alloy ········ Zhenxi Li, Hongwu Liu, Fan Gao, Chunxiao Cao 28

Localized Dissolution of Carbon Steel under Cathodic Polarization Conditions ········ Natalia Gladkikh, Marina Maleeva, Maxim Petrunin, Ludmila Maksaeva 29

Corrosion Resistance of UV Curable Hybrid Coatings on the Galvanized Sheet ········ Fengguo Liu, Wenju Tao, Yuzheng Wang, Youjian Yang, Xianwei Hu, Zhaowen Wang, Xiangxin Xue 30

Surface Modification of Aluminum in the Plasma of a Hollow-cathode Non-self-sustained Glow Discharge ········ Ivanov Yu. F., Tolkachev O. S., Denisov V. V., Petrikova E. A., Krysina O. V. 30

Characteristics of AA4045/AA3003 Clad Hollow Billet Prepared by Direct-Chill Process ········ Xing Han, Shuncheng Wang, Deng Nong, Kaihong Zheng, Haitao Zhang, Jianzhong Cui 31

Core-shell and Bi-phasic Structure Magnetic Nanoparticles for Biomedical Applications ········ Kamzin A.S., Phan M. H., Tkachenko M. V. 31

Preparation of TiO_2 Microfiltration Membrane on Porous Stainless Steel Tube
······ Shuai Li, Chao Zhang, Qinli Lv, Hua Zhang, Di He, Shumao Wang, Lijun Jiang 32
Микроструктурные Аспекты Хладноломкости Низколегированных Сталей
······ Кантор М.М., Воркачев К.Г., Солнцев К.А. 32
An Efficient Method for the Synthesis of SiBNC Preceramic Polymers with Different Si/B Atomic Ratios
······ Ji Xiaoyu, Shao Changwei, Wang Hao 33
Laser-interference Control of the Strain Field in the Process of Manufacturing Details
······ Kesariiskyi O.G., Kondrashchenko V.I. 33
EDC-induced Self-assembly of BSA-Au NCs ······ Wenli Zhu, Huili Li, Ajun Wan 34
Influence of Nanosecond Laser Radiation on the Surface Structure of Metals
······ R.R. Khasaya, Yu.V. Khomich, S.I. Mikolutskiy 34
Plasma Discharge Electrolysis of Nano Cuprous Oxide and Copper Particle in Solution ······ Lingling Shen 35
Low-temperature Curing High-silica Material with Increased Corrosion Resistance
······ Klimenko N.N., Levina Yu.N., Mikhailenko N.Yu. 35
C103 铌合金表面 Si-Cr-Ti 熔烧硅化物涂层 1400℃高温抗氧化性能研究
······ 汪 欣，李争显，杜继红，李晴宇 36
Oxidation Behavior of Slurry Fused Si-Cr-Ti Silicide Coating on C103 Alloy at 1400℃
······ Xin Wang, Zhengxian Li, Jihong Du, Qingyu Li 36
Structural Designing of High-coercivity (Nd, Pr, Dy, Tb)-Fe-B Magnets via $REMH_2$ Hydride Adding
······ Kolchugina N.B., Burkhanov G.S.,
Lukin A.A., Koshkid'ko Yu.S., Skotnicova K., Cwik J., Rogacki K., Drulis H. 37
Preparation and Photocatalytic Activity of TiO_2 Powder Codopd with La and Graphene Oxide
······ Wei Wang, Limei Yang, Songtao Huang, Zheng Xu, Zhen Cheng, Lu Jia 37
Phase Transformation and Morphology in the Process of Sythesis of YAG:Nd
······ Kolomiets T.Yu., Telnova G.B., Solntsev K.A. 38
Additive Manufacturing of Strong and Ductile Cu-15Ni-8Sn ······ Chao Chen, Gengming Zhang, Kechao Zhou 38
Model of Concrete Macrostructure
······ Kondrashchenko V.I., JING Guoqing, Kondrashchenko E.V., WANG Chuang 39
Comparative Study on Activity of Zn-Based Multicomponent Liquid Alloys between Different Thermodynamic Models ······ Heng Dai, Dongping Tao 39
Evaluation of Workability of Machine Parts on the Basis of the Odinga Criterion
······ Leonid Kondratenko, Viktor Terekhov, Lyubov Mironova 40

Extension of Molecular Interaction Volume Model to Electrolyte Solution ······· Congyu Zhang, Dongping Tao 40

Shock Waves in a Zinc Single Crystal
······························ Krivosheina M.N., Kobenko S.V., Tuch E.V., Lotkov A.I., Kashin O.A. 40

HfC 陶瓷先驱体的制备及其性能研究 ······························ 张丽艳，王小宙，王亦菲 41

Preparation and Properties of HfC Ceramic Precursor ················ Liyan Zhang, Xiaozhou Wang, Yifei Wang 41

Phase Formation and Ionic Conductivity of Zirconia-based Crystals Grown by Skull Melting Technique
······························ A.V. Kulebyakin, M.A. Borik, S.I. Bredikhin, V.T. Bublik, I.E. Kuritsyna,
E.E. Lomonova, F.O. Milovich, V.A. Myzina, V.V. Osiko, P.A. Ryabochkina, N.Yu. Tabachkova 42

Application of Molecular Interaction Volume Model to Fe-Based Liquid Alloys
······························ Yan Zhao, Heng Dai, Dongping Tao 42

Electrophysical Properties of the Lithium-Lanthanum Titanate Synthesized by Sol-gel Method
······························ Kunshina G.B., Efremov V.V., Ivanenko V.I. 43

Silicon-Glycerol-Polysaccharide Systems in Creating Topical Application Drugs
······························ Elena Yu. Larchenko, Elena V. Shadrina, Marina V. Ryaposova, Dmitry M. Kadochnikov,
Maria N. Isakova, Polina B. Zhovtyak, Sergey S. Grigoryev, Tat'yana G. Khonina 43

Controllable Modification of Glassy Composites with Ion-Exchange Technique
······························ A. A. Lipovskii, A. V. Redkov, A.A. Rtischeva, V. V. Rusan, D. K. Tagantsev, V. V. Zhurikhina 44

List of Articles for Publication ······························ 44

Structure and Corrosion Resistance of Ti-0.16 Pd Alloy after Equal Channel Angular Pressing
······························ Lotkov A.I., Kopylov V.I., Latushkina S.Yu., Grishkov V. N.,
Baturin A.A., Korshunov A.V., Abramova P.V., Girsova N.V., Timkin V.N., Zhapova D.Yu. 44

Structure and Properties of Self-expanding Intravascular NiTi Stents Doped with Si Ions
······························ Lotkov A.I., Kashin O.A., Kudryashov A.N., Krukovskii K.V. 45

Influence of the Jahn-Teller Effect on the Structure of Ferroelectrics, Magnets and Multiferroics
······························ Makarov Valery Nikolayevich, Kanygina Ol'ga Nikolayevna 45

Bioactive Materials for Restoration Surgery of Bone Tissue ················ Medkov M.A., Grishchenko D.N. 45

Directional Evolution of the Structure and Properties of HTSC Tapes under the Influence of Impacts Impulses
······························ Mikhailov B.P., Mikhailova A.B., Borovitskaya I.V., Burkhanov G.S., V.Ya.Nikulin, P.V.Silin 46

Annealing Temperature Effect on Protonic Conductivity of Aquivion Like Electrolyte Membranes
······························ Kamila Mugtasimova, Alexey Melnikov, Elena Galitskaya, Alexander Sivak, Vitaly Sinitsyn 46

Corrosion Resistance of Biocompatible Layered Composite Materials with Shape Memory Effect in Modeling
Media ············ Nasakina E.O., Baikin A.S., Kaplan M.A., Konushkin S.V., Sergienko K.V., Kovaleva E.D.,
Kolmakova A.A., Kargin Yu.F., Demin K.Yu., Sevost'yanov M.A., Kolmakov A.G., Simakov S.V. 47

The Technology of Preparation of Decellularized Human Liver Tissue Fragments to Create Cell - and Tissue-Engineered Liver Constructs
············· Nemets E.A., Kirsanova L.A., Basok Ju.B., Lymareva M.V., Schagidulin M.Ju., Sevastianov V.I. 47

The Use of Spark Plasma Sintering Method for High-rate Diffusion Welding of Ultrafine-grained α-titanium Alloys with High Strength and Corrosive Resistance
·····························Nokhrin A.V., Chuvil'deev V.N., Boldin M.S., Piskunov A.V., Kozlova N.A., Chegurov M.K., Popov A.A., Lantcev E.A., Kopylov V.I., Tabachkova N.Yu. 48

Studies Into the Impact of Mechanical Activation Modes on Optimal Solid-phase Sintering Temperature of Nano-and Ultrafine-grained Heavy Tungsten Alloys ························ Nokhrin A.V., Chuvil'deev V.N., Boldin M.S., Sakharov N.V., Baranov G.V., Popov A.A., Lantcev E.A., Belov V.Yu. 48

Spark Plasma Sintering of Bulk Ultrafine-grained Tungsten Carbide with High Hardness and Fracture Toughness··························· Nokhrin A.V., Chuvil'deev V.N., Blagoveshchenskiy Yu.V., Boldin M.S., Sakharov N.V., Isaeva N.V., Popov A.A., Lantcev E.A., Belkin O.A., S.P. Stepanov 48

Ammonolysis of Magnesiothermic Niobium and Tantalum Powders
······························· V. M. Orlov, R. N. Osaulenko, D.V. Lobov 49

Preparation and Properties of Nanostructured Materials Based on Polyphenoxazine and Bimetallic Co-Fe Nanoparticles ··························· Ozkan S.Zh., Karpacheva G.P. 49

Synthesis Methods of Hybrid Magnetic Materials Based on Polyphenoxazine and Fe_3O_4 Nanoparticles
······························· Ozkan S.Zh., Karpacheva G.P. 50

Comparative Kinetics of Extraction of Molybden-containing Components from the Waste Hydrotreatment Catalyst by Mineral Acids ·····················S.P. Perekhoda, E.Yu. Nevskaya, O.A. Egorova 50

Feathers of Formation of Vinul Siloxane Functional Nanoiayers on Zinc Surface
····················· Maxim Petrunin, Natalia Gladkikh, Ludmila Maksaeva, Marina Maleeva, Elena Terekhova 51

Action of Random Signal on Oscillatory Circuit with Ferroelectric Negative Capacitance
······························· A.A. Potapov, A.E. Rassadin, A.A. Tronov 51

On Solution of Fokker-planck-kolmogorov Equation for a Ferroelectric Capacitor with a Negative Capacitance by Means of the Krasovsky Series Expansion Method
······························A.A. Potapov, I.V. Rakut, A.E. Rassadin, A.A. Tronov 52

On Transient Response of *p-n* Junction on Some Ultrawideband Signals
······························A.A. Potapov, A.E. Rassadin, A.V. Stepanov, A.A. Tronov 52

Generators of Chaotic Electrical Oscillations on Basis of Ferroelectric Capacitor with a Negative Capacitance
······························· A.A. Potapov, A.E. Rassadin, A.A. Tronov 52

Nonlinear Dynamics of Fractals with Cylindrical Generatrix on Surface of Solid State

……………………………………………A.A. Potapov, A.E. Rassadin, A.V. Stepanov, A.A. Tronov 53

Investigation of Stability of Solitary Charge Wave in Infinite Transmission Line with Ferroelectric Capacitors with a Negative Capacitance ……………………………… A.A. Potapov, I.V. Rakut, A.E. Rassadin 53

Fractal Radioelement's, Devices and Fractal Systems for Radar and Telecommunications ……………………………………………Potapov A.A., Potapov Alexey A., Potapov V.A. 53

Magnetoresistance and Strength Properties of HTSC Composite Tape Joints Obtained by Soldering ………………………………… Prosvirnin D.V., Troitskii A.V., Markelov A.V., Demikhov T.E., Antonova L.Kh, Mikhailov B.P., Molodyk A.A., Mikhailova G.N. 54

Modes for Ceramics Based on Aluminum Oxynitride, and Their Influence on the Properties of the Material ……………………………………………… Prosvirnin D.V., Kolmakov A.G., Alikhanyan A.S., Samokhin A. V., Antipov V.I., Larionov M. D., Titov D.D. 54

Investigation of the Surface of Destruction of Aluminum Coatings at Negative Temperatures ……………………………………… Prygaev A., Vyshegorodtseva G., Buklakov A., Morozova V. 54

Use of Functional Properties of Nithinol Alloys for Increasing the Energy Efficiency of Pumping Equipment ……………………………………………………Prygaev Alexander K., Dubinov Yuri S., Dubinova Olga B., Fatkhutdinov Ruslan R., Nakonechnaya Ksenia V. 55

From Nano-to Femtotechnology and Back. About Possibility of Obtaining a Neutron Substance in Laboratory Conditions······ Ryazantsev G.B., Beckman I.N., Lavrenchenko G.K., Khaskov M.A., Pokotilovskiy Yu. N. 55

Control of Powders Properties under Conditions of Synthesis and Treatment in DC Thermal Plasma Jet ………………………………… Samokhin A. V., Alekseev N. V., Kirpichev D. E., Astashov A. G., Fadeev A. A., Sinaiskiy M. A., Tsvetkov Y. V. 56

Calculation of Direct and Inverse Current-voltage Characteristics of Schottky Barrier Height and on the Border of Metal-semiconductor in a Nonlinear Model for Diodes Based on SiC ………………………………… A.V. Sankin, V.I. Altukhov, A.S. Sigov, S.V. Filipova, E. G. Janukjan 56

Formation of Composite Materials Depending on the Geometrical Parameters ············Seregin A.V., Nasakina E.O., Sudarchikova M.A., Sprygin G.S., Khimyuk Ya.Ya., Demin K.Yu., Kaplan M.A., Konushkin S.V., Fedyuk I.M., Yakubov A.D., Sevost'yanov M.A., Kolmakov A.G. 57

Technology of Tissue Engineering and Regenerative Medicine in the Treatment of Damaged Articular Cartilage ····················V.I. Sevastianov, Yu.B. Basok, A.M. Grigoryev, L.A. Kirsanova, V.N. Vasilets, S.V. Gautier 57

Simulation of Stress-strain State of Pipeline Parts under Plastic Deformation and Fracture········ Shabalov I.P., Velikodnev V.Y., Murzakhanov G.Kh., Kalenskii V.S., Nastich S.U., Barsukov A.A., Arsenkin A.M. 57

Promising Ceramic Materials Based on Elementoxane Precursors ······Shcherbakova G.I., Varfolomeev M.S.,

Krivtsova N.S., Storozhenko P.A., Moiseev V.S., Yurkov G.Yu., Ashmarin A.A. 58

Research of Thermophysical Properties of Organoplastics Based on Polyarylate Sulfone Block Copolymer
………………………… Shustov G. B., Burya A. I., Grashchenkova M. A., Shetov R. A. 58

Low-temperature Extraction-pyrolytic Synthesis of Nano-size Composites Fased on Metal Oxides
…………………………N.I. Steblevskaya, M.A. Medkov, M.V. Belobeletskaya 59

Electrochemical Forming of Electroconductive Poly-Fe (III)- Aminophenylporphyrin Films in Various Solvents
………………… Tesakova M.V., Kuzmin S.M., Chulovskaya S.A., Semeikin A.S., Parfenyuk V.I. 59

Role of Graphene Sheets in Formation of Metal Oxides Based Hybrid Nanostructures
………………… Trusova E.A., Kirichenko A.N., Kotsareva K.V., Polyakov S.N., Abramchuk S.S. 59

Effect of the Power Pulses of Deuterium Plasma on the Structure and Mechanical Properties of V-Ga Alloys
…………………………A.B. Tsepelev, V.F. Shamray, V.P. Sirotinkin, N.A. Vinogradova 60

Highly Effective Activated Carbon from Wood for Supercapacitors: Synthesis and Research
………………………… Vervikishko D.E., Kochanova S.A., Kiselyova E.A., Shkolnikov E.I. 60

Recent Advances in Laser-Pulse Melting of Graphite at High Pressure……………… P.Vervikishko, M. Sheindlin 60

The Influence of Aluminium Addition on the Structure and Magnetic Properties of the Pseudobinary $Sm_2(Fe_{1-x}Al_x)_{17}$ Alloys and Their Hydrides
…………………………Veselova S.V., Verbetsky V.N., Savchenko A.G., Shchetinin I.V. 61

ВЛИЯНИЕ АЛЮМИНИЯ НА СТРУКТУРУ И МАГНИТНЫЕ СВОЙСТВА ПСЕВДОБИНАРНЫХ СПЛАВОВ $Sm_2(Fe_{1-x}Al_x)_{17}$ И ИХ ГИДРИДОВ
…………………………Веселова С.В., Вербецкий В.Н., Савченко А.Г., Щетинин И.В. 62

Elemental Analysis of Ceramic Materials of Medical Applying
………………………… Volchenkova V.A., Kazenas E.K., Andreeva N.A., Ovchinnikova O.A., Paunov A.K., Penkina T.N., Rodionova S.K., Fomina A.A., Podzorova L.I., Ilyicheva A.A. 63

Potential Output from Technological Deadlock in Creation of New Generation Technique……… Georg Volkov 63

Microstructure Investigation of Low Carbon Low Alloy Steel Suitable for Application in Arctic Constructions
…………………………Vorkachev K.G., Kantor M.M., Solntsev K.A. 63

Investigation of Wear Resistance of Aluminum and Zinc Coatings at Low Temperatures
…………………………Vyshegorodtseva G., Buklakov A., Morozova V., Vyshegorodtseva I. 64

The New Nanocomposite Carbon Material for High-field Electron Sources
…………………………Yafarov R.K., Shanygin V.Ya., Nefedov D.V. 64

Metal Nanowires-New Type of Nanomaterial: Fabrication by Matrix Synthesis Technique and Investigation of Structure and Properties ………………………… D.L. Zagorskiy,

S.A. Bedin, I.M.Doludenko, G.G.Bondarenko, K.V. Frolov, V.V.Artemov, M.A. Chuev, A.A.Lomov 65

Membrane Palladium-Based Alloys for High Purity Hydrogen Production

······················ G.S. Burkhanov, N.R. Roshan, E.M. Chistov, T.V. Chistova, S.V. Gorbunov, A.D. Zakharov 65

Polarization of Glass with Positive and Negative Charge Carriers

······················ V. V. Zhurikhina, M. I. Petrov, A. A. Rtischeva, A. A. Lipovskii 66

Laser Melting and Thermal Treatment of Co-Cr Based Alloys

······················ Drápala J., Losertová M., Konečná K., Kostiuková G. 66

Thermal Effects in Composites Hydroxyapatite-polysaccharide ·········· Svetlana A. Gerk, Olga A. Golovanova 66

Crystallization of Carbonated-hydroxyapatite and Si-hydroxyapatite on Titanium Implants

······················ Olga A. Golovanova 67

Nanostructuring of the Bismuth Single Crystal Surface (111) under the Action of Atomic Hydrogen

······················ O.I. Markov, Yu.V. Khripunov 67

Creation of Competitive Composite Materials in the Presence of Interphase Layer on the Border of the Inclusion and the Matrix ······················ Pavlov S. P., Bodyagina K. S. 68

Study on the Microstructure and Properties of Ag-SnO_2 and Ag-CuO-La_2O_3 Electrical Contact Materials Prepared by Powder Metallurgy ······················ Xie Ming, Wang Song, Zhang Jiming, Li Aikun, Hu Jieqiong, Wang Saibei, Yang Youcai, Chen Yongtai, Liu Manmen 68

Hydrogen Effects on Different Properties of Biocompatible Metallic Materials ······················ Losertová M. 69

Research of Reacting System Self-Purification in the Process of Self-Propagating High-Temperature Synthesis (SHS) ·········· A.A. Potekhin, D.A.Gorkaev, A.Yu. Postmikov, A.I. Tarasova, A.Ya. Malyshev 70

The Use of Analysis Spectral Methods to Assess Powder Compositions for Self-Propagating High-Temperature Synthesis ······················ A. Yu. Postnikov, V.V. Mokrushin, A.A. Potekhin, I.A. Tsareva, O.Yu. Yunchina, M.V. Tsarev, D.V. Chulkov, P.G. Berezhko 70

Study of Phase Diagram and Thermodynamic Parameter of Au-Sn-Pt System ············ Xie Ming, Hu Jieqiong, Zhang Jiming, Wang Saibei, Li Aikun, Liu Manmen, Yang Youcai, Chen Yongtai, Fang Jiheng 71

Application of Metal Hydrides as Pore-Forming Agents for Obtaining Metal Foams ············ N.V. Anfilov, A.A. Kuznetsov, P.G. Berezhko, A.I. Tarasova, I.A. Tsareva, V.V. Mokrushin, M.V. Tsarev, I.L. Malkov 71

Application of Acoustic Emission Method to Study Metallic Titanium Hydrogenation Process

······················ A.A. Kuznetsov, P.G. Berezhko, S.M. Kunavin, E.V. Zhilkin, M.V. Tsarev, V.V. Yaroshenko, V.V. Mokrushin, O.Yu. Yunchina, S.A. Mityashin 72

Device for Reversible Hydrogen Isotope Storage with Aluminum Oxide Ceramic Case

······················ I.P. Maximkin, A.A. Yukhimchuk, V.V. Baluev, I.L. Malkov, R.K. Musyaev, D.T. Sitdikov, A.V. Buchirin, V.V. Tikhonov 72

The Studies on the Interfacial Evolution Behavior of $ZrO_2(ZrB_2)$ Active Diffusion Barrier

Lintao Liu, Zhengxian Li, Yunfen Chen

(School of Metallurgical Engineering, Xi'an University of Architecture and Technology, Xi'an 710055, P.R. China)

ABSTRACT: In this paper, two sample groups, N5/(ZrB_2+ZrO_2)/NiCrAl and N5/ZrO_2/NiCrAl, were prepared on Ni-based single crystal alloy (Rene N5) substrate by electron beam physical vapor deposition (EB-PVD). Both sample groups were exposed to isothermal oxidation at 900℃ for 5 h and at 1000℃, for 250 h, 300 h or 350 h. The microstructural evolution and deterioration failure behavior was investigated by scanning electron microscopy (SEM) and energy dispersive spectroscopy (EDS). The results suggest that the introduction of ZrB_2 decelerates the interfacial reaction rate of the active diffusion barrier of Al_2O_3 but did not affect the final formation of the Al_2O_3 diffusion barrier with anti-diffusion properties. Moreover, the introduction of ZrB_2 prolongs the service life of active diffusion barrier structure and changes its failure mode.

KEY WORDS: MCrAlY coating; diffusion barrier; interfacial evolution

Mathematical Modeling and Computer Simulation of Steelmaking Technologies

Grigorovich K.V., Komolova O.A., Gorkusha D.V.

(Baikov Institute of Metallurgy and Material Science, RAS, Leninskii pr., 49, Moscow Russia, 119991)

ABSTRACT: In this study the mathematical models, algorithms and software for dynamic simulation of steelmaking technology in ladle furnace and vacuum degasser were developed and created a new method of technology optimization. The inclusion control procedures using the fractional gas analysis method to increase the steel cleanliness, to detect and correct of sources of inclusions origin during the ladle treatment were developed. The methods of quantitative metallography, X-ray microprobe and fractional gas analysis for the oxide inclusions control and their quantity determination in the probes sampled from the ladle furnace, ladle vacuum degasser and tundish during the ladle treatment and casting of steel were applied. It was demonstrated that ladle treatment technologies of different steels grades can be optimized using combination of software designed and fractional gas analysis method to provide the favorable composition of oxide inclusions and to increase the steels quality.

Development of Molybdenum-based Metallic Glasses

Gao Xuanqiao, Li Bin, Zhang Wen, Ren Guangpeng, Xue Jianrong, Li Laiping

(Northwest Institute for Nonferrous Metal Research, Xi'an 710016, Shaanxi, China)

ABSTRACT: Amorphous alloys have many unique properties on account of the characteristics of both metal and glass. Due to the excellent thermal stability and ultra-high hardness, Mo-based metallic glasses have extensive application prospect, which has attracted much attention. In this paper, the development history and research progress of Mo-based amorphous alloys are summarized from the aspects of composition system, preparation method, microstructure and physical and mechanical properties. Furthermore, the existing problems are briefly described, and the future research direction is prospected.

KEY WORDS: metallic glass; molybdenum alloy; preparation; mechanical properties

Investigations of Decarburisation Rates of High Chromium Iron Base Melts by Low-temperature Plasma. Mathematical Model and Experiments

K. V. Grigorovich[1], O. A. Komolova[2], B. A. Rumyantsev[1]

(1. Baikov Institute of Metallurgy and Materials Science, Russian Academy of Sciences, Moscow;
2. National Investigation Technology University "MISIS", Moscow)

ABSTRACT: The process of interaction of low temperature oxygen content plasma with iron-based melt was simulated in laboratory experiments. The influence of the power of the plasma arc on the evaporation rate and surface temperature of the melt was investigated. The rate constants of alloy components evaporation are determined. It was found that the alloys considered, Fe-Cr-Ni melt is characterized by the highest evaporation rate in the neutral atmosphere of a laboratory plasma furnace.

The treatment of Cr and Cr-Ni iron base alloys by oxidative plasma was investigated by experimentally, in order to analyze the carbon oxidation processes within the small reaction region of the metal and the plasma flare.

It was found, that deep decarburization of high-chromium melt by plasma containing no more than 14% oxygen is possible, without loss of the alloying elements, especially chromium.

The mathematical model describes the interaction of the metal melt and low-temperature plasma was developed. Comparison shows that the model data and experimental results are in good agreement.

Advances of Mo-Re Alloys on Microstructures and Mechanical Properties

Li Laiping, Gao Xuanqiao, Hu Zhongwu, Liang Jing, Lin Xiaohui, Xue Jianrong

(*Northwest Institute for Nonferrous Metal Research, Xi'an 710016, China*)

ABSTRACT: This paper reviews the recent research and development of Mo-Re alloys that are excellent structural materials for high temperature application with high strength at elevated temperature, good stability and compatibility with nuclear fuel and alkali liquid metal. These alloys are manufactured into various equipments for chemical, aerospace and nuclear industry. Up to date, the microstructures and mechanical properties of Mo-Re alloys have been studied before and after neutron irradiation. This paper first reviews the microstructure evolution of Mo-Re alloys induced by neutron irradiation condition, and mechanical properties such as radiation hardening and embrittlement are then discussed.

KEY WORDS: refractory alloy; Mo-Re alloy; neutron irradiation; microstructure; mechanical properties

Development of Methods for Preparation of Oxide Ceramic Systems via Individual Precursors

V.M. Novotortsev, Zh.V. Dobrokhotova, A.V. Gavrikov, P.S. Koroteev, N.N. Efimov

(*IGIC RAS, Moscow, Russia*)

Effect of Multilayered Structure on the Properties of Ti/TiN Coating

Yanfeng Wang, Zhengxian Li, Jihong Du, Haonan Wang, Changwei Zhang

(*Northwest Institute for Nonferrous Metal Research, Xi'an 710016, China*)

ABSTRACT: Metal nitride hard coatings, such as TiN and TiAlN, were widely used to protect materials because of their higher hardness and wearing properties. However, these coatings always contain a high degree of internal stress which could arouse adhesion problems. It's hard to synthesize monolayer TiN or TiAlN coatings thicker than 10μm by PVD method. The multilayer composite structure, offering an efficient way of controlling residual stress, is an effective way to synthesize thicker hard coatings. In this study, a series of multilayer composite Ti/TiN coatings with different composite periods were synthesized by plasma enhanced ion plating and how the multilayered structure affects the coating's mechanical properties was studied in detail. The result shows that the coating's mechanical properties are strengthened as increasing the periods of composite Ti/TiN layer. The micro hardness can reach to about 2750HV0.25 albeit possessing better toughness and higher thickness (i.e.>50μm). And also the coating's tribological performance is improved with a lower dry friction coefficient (about 0.35) and higher wearing resistance by alternating 48 periods of Ti/TiN layer. However, the coating's bonding strength is weakened as increasing the period of Ti/TiN layer blindly due to the weak interface of coating and substrate.

KEY WORDS: Ti/TiN coating; PVD; multilayered structure; micro hardness; tribological performance

Synthesis and Investigation of the $Cd_{1-x}Fe_xCr_2S_4$ Solid Solutions as a Prospective Basis for the Magnetoelectrical and Magnetocapacitive Materials

V.M. Novotortsev, T.G. Aminov, G.G. Shabunina, N.N. Efimov

(*IGIC RAS, Moscow, Russia*)

Current Research in Silver Nanowires

Yongqian Wu, Kunkun Chen, Bosheng Zhang, Qigao Cao
(Northwest Institute of Nonferrous Metal Research, Xi'an 710016, China)

ABSTRACT: Silver nanowires (AgNWs) have attracted considerable attention because of their excellent conductivity, thermal conductivity, flexibility and unique size effect of nano materials. AgNWs are deemed to be a leading candidate material to replace ITO as a result of they have excellent photoelectric performance and mechanical performance. Various synthesis of silver nanowires are introduced in this paper.

Comparative Kinetics of Extraction of Molybden-containing Components from the Waste Hydrotreatment Catalyst by Mineral Acids

S.P. Perekhoda[1], E.Yu. Nevskaya[2], O.A. Egorova[2]
(1. Baikov Institute of Metallurgy and Materials Science, RAS, Moscow;
2. Peoples' Friendship University of Russia. RUDN University, Moscow)

Fabrication of Ag Sheathed FeSe Superconducting Tapes

Shengnan Zhang[1], Chengshan Li[1], Jianqing Feng[1], Jixing Liu[1,2], Botao Shao[1,3], Pingxiang Zhang[1]

(*1. Superconducting Materials Research Center, Northwest Institute for Non-Ferrous Metal Research, Xi'an 710016, China*; *2. School of Materials Science and Engineering, Northeastern University, Shenyang 110016, China*; *3. School of Materials Science and Engineering, Xi'an University of Technology, Xi'an 710048, China*)

ABSTRACT: FeSe superconducting tapes with different metal sheaths of Nb/Cu composite, Ag and Fe have been fabricated with traditional powder in tube (PIT) process, respectively. With the same cold working process and heat treatment, the influences of different sheath materials on the phase composition and microstructures of FeSe filaments have been analyzed. Due to the reaction between Nb-Se and Ag-Se, Nb_2Se and Ag_2Se particles can be clearly observed embedded in the filaments. Therefore, it is necessary to introduce a new method for the fabrication of Ag or Nb sheath FeSe wires in order to avoid the introduction of Fe in magnet system during practical applications. Based on our previous study, high energy ball milling process has been performed to achieve precursor powders with amorphous Fe-Se binary compound instead of elemental Se. The formation of Ag-Se compounds has been successfully avoided and the superconducting tapes with high superconducting FeSe critical temperature of 9.0 K have been obtained.

KEY WORDS: superconductors; FeSe; powder in tube process; microstructure

Synthesis and Optical Properties of Aluminum Oxynitride (Alon) Doped with Rare-earth Metals Ions EU^{2+} and CE^{3+}

N.S. Akhmadullina[1], A.S. Lysenkov[1], V.V. Yagodin[2], A.V. Ishchenko[2], Yu.F. Kargin[1], B.V. Shul'gin[2]

(*1. A.A. Baikov Institute of Metallurgy and Material Science of Russian Academy of Sciences, Leninsky prospect, 49, 119991, Moscow, Russia*; *2. B.N. El'tsin Ural Federal University, Mira st., 19, 620002, Ekaterinburg, Russia*)

Preparation and Mechanical Properties of SiC Whisker Reinforced Multi-phase Mo-Si-B Alloy

Bin Li[1], Laiping Li[1], Xiaohui Lin[1], Huiqing Fan[2], Pingxiang Zhang[1]

(1. Northwest Institute for Non-ferrous Metal Research, Xi'an 710016, China; 2. State Key Laboratory of Solidification Processing, Northwestern Polytechnical University, Xi'an 710072, China)

ABSTRACT: Fine-grained and multi-phase Mo-Si-B alloy because of containing high amounts of grain/phase interfaces can be optimized for performance improvement by modifying the interface. Thus, this study basing on structure design attempted to introduce the nano-scale whisker in the multi-phase Mo-Si-B alloy. Microstructure, composition, strength and toughness were experimentally examined and solid-state reactions basing on thermodynamic calculation were analysed to evaluate the effect of SiC whisker. The microstructure of Mo-12Si-8.5B-SiC alloy exhibited a fine-grained intermetallic matrix in which some isolated α-Mo particles are distributed, which is different from this of Mo-12Si-8.5B alloy that exhibited some silicide particles bonded by a continuous α-Mo matrix. Some SiO_2 particles also distribute at the grain or phase boundaries of the alloy, but the amounts can be decreased by the SiC addition which is mainly related to the grain/phase boundary purifying effect derived from the interface reaction of SiC whisker and Mo matrix. The SiC whisker can enhance the strengths and fracture toughness of the Mo-12Si-8.5B alloy, due to the multi-phase microstructure and low oxide impurity concentrations.

KEY WORDS: Mo-Si-B alloy; multi-phase; whisker; reaction addition; mechanical property

Magnetic Properties of the Hard Magnetic Alloy of FE-27CR-10CO-2MO

Alymov M. I., Milyaev I.M., Yusupov V. S., Zelensky V.A., Milyaev A.I., Ankudinov A.B., Abashev D. M.

(Institution of Russian Academy of Sciences A.A. Baikov Institute of Metallurgy and Material Science RAS, Russia)

Effect of Strain Rate on Texture Evolution of TLM Titanium Alloy During Hot Deformation

X.F. Bai, Y.Q. Zhao, X.M. Wang, J.Y. Liu

(*Northwest Institute for Nonferrous Metal Research, Xi'an 710016, China*)

ABSTRACT: In this work, the texture evolution of a near β Ti-3Zr-2Sn-3Mo-25Nb (TLM) biomedical titanium alloy in the hot compression of different strain rates from $0.001s^{-1}$ to $1s^{-1}$ has been investigated. The XRD examination shows that samples of hot deformation consist of β phase only. The development of texture has been explained in terms of orientation distribution functions (ODFs) of α and γ fibres. The results show that the texture changes with the increase of strain rate. $0.1s^{-1}$ strain rate is favorable to the orientations of {001}<110> and {112}<110>, while the orientation between {001}<110> and {112}<110> is main texture at the strain rates of $0.001s^{-1}$ and $1s^{-1}$. Specifically, compared with $0.001s^{-1}$ strain rate, the orientation density of texture components increase at the condition of $1s^{-1}$.

KEY WORDS: TLM titanium alloy; texture; hot compression; strain rate

Microstructure of the Polycrystalline Alloys of the $Al_{85}Ni_{11-x}Fe_xLa_4$ System

N.D. Bakhteeva[1], N.N.Kolobylina[2], E.V.Todorova[1], A.G.Ivanova[3], A.L. Vasiliev[2]

(*1. Baikov Institute of Metallurgy and Materials Science, Russia, nbach@imet.ac.ru; 2. National Research Centre "Kurchatov Institute", Russia; 3. Shubnikov Institute of Crystallography of FSRC "Crystallography and Photonics" RAS, Russia*)

Preparation and Properties of Foam Titanium for Biological Implants

Li Zengfeng, Tang Huiping, Chen Gang, Tan Ping, Zhao Shaoyang, Shen Lei

(*State Key Laboratory of Porous Metal Materials, Northwest Institute for Nonferrous Metal Research, Xi'an 710016, China*)

ABSTRACT: In this paper, titanium foam was prepared by a space-holder method. Ammonium bicarbonate was chosen as the pore-forming agent. Titanium powders and TiH_2 powders were selected as raw powders. In this thesis, the control mechanism and optimal process of pore structure were investigated. The results showed that the compressive yield strength and young's modulus of titanium foam decreased as sintering temperature and sintering time increased; with increased pore-forming agent content, foam titanium compressive yield strength and young's modulus decrease. Based on the studies mentioned above, the porosity of titanium foam, which was produced with TiH_2 powders as the raw material by adding different concentrations of pore-forming agent, ranges from 48% to 77%. Pore size is controlled by screening the pore-forming agents. Titanium foam was successfully prepared whose macro pore size varies from 300 to 500 μm and keyhole size was in micrometer scale.

KEY WORDS: titanium foam; pore-forming agent; porosity; compression yield strength; young's modulus

Fiber-optical Acoustic Emission Sensors for Registering Damage in Polymeric Composite Materials

Bashkov O.V., Romashko R.V., Zaikov V.I., Khon H., Bashkov I.O., Bryansky A.A.

(*Komsomolsk-on-Amur State Technical University, Russia*)

ABSTRACT: Features of recording acoustic emission waves by fiber-optic sensors are shown. Fiber-optic sensors were constructed according to the scheme of an adaptive interferometer, which uses the formation of dynamic holograms. Optical fibers of the interferometer are ultrasonic wave sensors. The sensors were mounted in the polymer composite material during the sample manufacturing process. The test sample was a plate of a multilayer fiberglass composite material. Studies have shown that the sensitivity of the interferometer makes it possible to register the acoustic emission signals that were excited by the Hsu-Nielsen source. The anisotropy of the composite material properties showed a different sound velocity in different directions of the plate. The sound velocity, measured by piezoelectric transducers and fiber-optic sensors, is different. The difference of the sound velocities associated with the registration feature group Lamb waves in the plate. To obtain an informative component of the acoustic emission signal, a wavelet transform was used.

KEY WORDS: acoustic emission; fiber optic sensor; piezoelectric transducer; adaptive interferometer; photorefractive crystal; Su-Nielsen source; polymer composite material; lamb wave; group velocity; wavelet

Effects of Heattreatments on the Properties of Coarse-grained WC-10 Cocemented Carbides with Low Carbon Content

X.C. Xie, Z. W. Li, R. J. Cao, Z. K. Lin

(*Powder Metallurgy and Special Materials Department, General Research Institute for Nonferrous Metals, Beijing 100088, China*)

ABSTRACT: The effects of cryogenic treatment and aging tempering treatment on the properties of low carbon coarse-grained WC-10 Cocemented carbides were investigated. During the heat treatments, the nanoscaled particles and compressive stress have effect on the phase transformation, structure, residual stress, hardness and wear resistance.

KEY WORDS: cryogenic; aging tempering; low carbon;coarse-grained; cemented carbide

Acoustic Emission of Fatigue Damage in Structured Pure Titanium VT1-0

Bashkov O.V., Bashkova T.I., Popkova A.A., Sharkeev Yu.P., Eroshenko A.Yu., Tolmachev A.I.

(*Komsomolsk-on-Amur State Technical University, Russia*)

ABSTRACT: The results of studying the kinetics of accumulation of fatigue damages of the titanium VT1-0 in a different structural state by the acoustic emission method are presented. The different structural state of titanium is obtained by the equal-channel angular pressing. By the nature of the accumulation of the total AE, the stages of fatigue are allocated. A later registration of AE signals with a decrease in the grain size was noted. The large-crystal structure is characterized by an earlier detection of the dislocation-type AE signals.

KEY WORDS: acoustic emission (AE); equal channel angular pressing; microcracks; dislocations; fatigue; titanium; submicrocrystalline structure; ultrafine-grained structure; deformation

Film-forming Characteristic of Lead-silver-calcium Alloy Anode during Zinc Electrowinning

H.T. Yang[1], C.L. Fan[1], Y.G. Wei[2,3], B.M. Chen[2,3], Z.C. Guo[3,4]

(1. State Key Laboratory of Multi-Phase Complex Systems, Institute of Process Engineering, Chinese Academy of Sciences, P.O. Box 353, Beijing 100190, China; 2. State Key Laboratory of Complex Nonferrous Metal Resources Clean Utilization, Kunming University of Science and Technology, Kunming 650093, China; 3. Faculty of Metallurgical and Energy Engineering, Kunming University of Science and Technology, Kunming 650093, China; 4. Kunming Hendera Science and Technology Co., Ltd., Kunming 650106, China)

ABSTRACT: Film-forming characteristic of lead-silver-calcium alloy anode over 15d of galvanostatic electrolysis in acidic zinc sulfate electrolyte solution were investigated. Anodic polarization curves, cyclic voltammetry (CV) curves, quasi-stationary polarization curves (Tafel), and electrochemical impedance spectroscopy (EIS) were employed to study the electrochemical behaviours. The microscopic morphology and phase composition of the anodic oxide film were observed by scanning electron microscopy (SEM) and X-ray diffraction (XRD), respectively. The electrocatalytic activity and reaction kinetics of the anodes varied evidently during electrolysis, indicating the formation and stabilization of the anodic oxide film. With prolonged electrolysis, the potential and overpotential of the oxygen evolution in the anodes mainly exhibited a declining trend, contrary to the trend of the electrode surface exchange current density. This depolarization may have been caused by the increasing roughness of the anodic film and the catalytic effect of increasing α-PbO_2 content. With prolonged electrolysis time, the cathodic peak (β-PbO_2 and α-$PbO_2 \rightarrow PbSO_4$) mainly showed a rising trend and gradually moved toward the negative direction. A cauliflower-like and regular microscopic structure was also clearly observed. The corrosion phases of the anodic oxide film mainly consisted of $PbSO_4$, Pb, PbS_2O_3, α-PbO_2, and β-PbO_2. The peak intensity of $PbSO_4$ also exhibited a declining trend, contrary to that of α-PbO_2. β-PbO_2 was almost constant. $PbSO_4$ showed preferential growth orientation toward the (021), (121) and (212) planes, whereas α-PbO_2 showed preferential growth orientation toward the (111) plane.

KEY WORDS: film-forming characteristic; lead-silver-calcium alloy anode; electrochemical behavior; anodic oxide layer; zinc electrowinning

Hydrogen Effect on the Martensitic Transformations and the Superelasticity of TiNi Based Alloy with Ultrafine-grained Structure

Baturin A.A.[1,2], Lotkov A.I.[1], Grishkov V. N.[1], Rodionov I.S.[1]

(1. Institute of Strength Physics and Materials Science SB RAS, Russia, lotkov@ismps.tsc.ru; 2. Tomsk Polytechnic University, Russia)

ABSTRACT: The paper presents the results of a study the hydrogen effect on the structural-phase transformations and the superelasticity in binary ultrafine-grained (UFG) TiNi based alloy after diffusion redistribution hydrogen as a result of aging at room temperature. The redistribution of hydrogen in the process of long-term aging after electrolytic hydrogenation of UFG wire specimens the $Ti_{49.1}Ni_{50.9}$ (at.%) stabilizes the B2 structure. Superelasticity in samples aged at room temperature after hydrogenation noticeably becomes wors.

Selective Hydrolysis and Removal of Trace Titanium during the Chlorination Process for High Purity Vanadium Pentoxide

Chuanlin Fan, Haitao Yang, Qingshan Zhu

(State Key Laboratory of Multiphase Complex Systems, Institute of Process Engineering, Chinese Academy of Sciences, Beijing 100190, P. R. China)

ABSTRACT: Given the strong demands for high purity vanadium products dedicated for all-vanadium redox flow battery (VRFB) that is used for large-scale energy storage, numerous technologies for preparation of high purity vanadium pentoxide have been developed. Metallurgical grade vanadium oxide can be feasibly transformed to gaseous vanadium oxytrichloride using a carbochlorination process at 400-600 ℃, which is condensed to a liquid intermediate product with a boiling point of 127 ℃ and purified by distillation. And then high purity vanadium pentoxide is produced from the hydrolysis of obtained high purity vanadium oxytrichloride. In industrial practice, metallurgical grade vanadium oxide is mainly extracted from vanadium-titanium bearing magnetite. A small amount of titanium is inevitably into vanadium oxide stock, subsequently chlorinated to titanium tetrachloride and condensed with vanadium oxytrichloride. Trace titanium tetrachloride is difficult to be removed during the distillation, due to similar saturated vapor pressures of it and vanadium oxytrichloride, which becomes an important bottleneck for improvement of product purity. In present work, based on thermodynamic calculation and analysis, we found that titanium tetrachloride reacts with water significant preferentially to form corresponding oxide, compared with vanadium oxytrichloride. Then we proposed a novel method to remove trace titanium tetrachloride from vanadium oxytrichloride by selective hydrolysis – distillation. The related experiments were conducted to validate the feasibility of trace titanium removal using the novel process. And vanadium pentoxide with purity higher than 99.99 wt% (4N) was further prepared from the purified vanadium oxytrichloride by precipitation – calcination. The proposed method can effectively remove titanium impurity and increase the purity of vanadium pentoxide, which is significant for improving production efficiency of high purity vanadium pentoxide using the chlorination process.

KEY WORDS: high purity vanadium pentoxide; chlorination process; vanadium oxytrichloride; titanium tetrachloride; selective hydrolysis

Reference Complex for Metrological Ensuring of Ultrasonic Measurements in Solid Media

P.V. Bazylev, I.Y. Krumgolz, V.A. Lugovoy

(VNIIFTRI, Far Eastern Branch, Khabarovsk, Russia)

Study on Crack Propagation Rate of 600MPa High Strength Steel for Mn-Ti and Nb-V System

Li Xiaolin[1,2], Cui Yang[1,2], Yang Xiaohe[3], Jin Zhao[3]

(*1. Shougang Research Institute of Technology, Beijing 100043, China; 2. Beijing Key Laboratory of Green Recyclable Process for Iron & Steel Production Technology, Beijing 100043, China; 3. Shougang Jingtang Steel Company, Caofeidian, Hebei, 063200, China*)

ABSTRACT: The microstructure of 600MPa Mn-Ti and Nb-V steels for cold stamping axle housing was investigated by means of optical microscope. On the low-frequency fatigue tester MTS-810 with frequency of 10 Hz, the crack propagation rate of two chemical components of the tested steels was tested by three-point bending method. The relationship between the propagation rate da/dN and the amplitude of stress intensity factor ΔK was given. The results showed that the microstructure all of two tested steels were ferrite and pearlite, the grain size of Mn-Ti steel and Nb-V steel was 5.8μm and 3.3μm, respectively. The coefficient C and m in Paris equation of Mn-Ti steel and Nb-V steel were determined as m=2.6298, C=4.13×10^{-12}mm/cycle and m=2.6799, C=1.30×10^{-12}mm/cycle respectively, and the threshold value of crack propagation ΔK_{th} was 46.50 MPa·$m^{1/2}$ and 66.58 MPa·$m^{1/2}$, respectively.

KEY WORDS: cold stamping axle housing steel; fatigue crack; stress intensity factor amplitude; threshold value of crack propagation; grain size

Spark Plasma Sintering of High-strength Fine-grained Alumina Ceramics

Boldin M.S., Chuvil'deev V.N., Popov A.A., Nokhrin A.V.

(*Lobachevsky State University of Nizhny Novgorod, Russia*)

Effect of GH4169 Alloy Powder Characteristics on Selective Laser Melting Process

DU Kaiping[1,2,3], SHEN Jie[1,2,3], ZHANG Shuting[1,2,3], YU Yueguang[1,2,3]

(1. Beijing General Research Institute of Mining and Metallurgy, Beijing 100160, China; 2. Beijing Key Laboratory of Special Coating Materials and Technology, Beijing 102206, China; 3. Beijing Industrial Parts Surface Hardening and Repair Engineering Technology Research Center, Beijing 102206, China)

ABSTRACT: The product quality of selective laser melting (SLM) is closely related to the alloy powder characteristics. In this work, the GH4169 alloy powders with B mass concentrations of 0 and 0.005% were prepared with German vacuum atomization furnace and formed using SLM equipment. The effect of B on the preparation and SLM process of GH4169 alloy powder were also discussed. The results show that, with the same particle size distribution, the SLM sample with a B mass concentration of 0.005% has fewer holes, higher density and hardness compared with the SLM sample without B element. Therefore, adding B element could increase the median size of alloy powder applicable to SLM process from 26.33 μm to 37.34 μm. In addition, considering that the B element could improve the yielding rate of the above alloy powder from 9.3% to 20.8%, the alloy powder costs can be significantly reduced by 55.3%.

KEY WORDS: GH4169 alloy powder; vacuum gas atomization; selective laser melting; B element; particle size distribution

Investigation of the Processes Induced by High-dose Ar^+ Implantation into Vanadium and Vanadium-based Alloy V-4.51 Ga-5.66 Cr

I. V. Borovitskaya[1], S. N. Korshunov[2], A. N. Mansurova[2], N. A. Vinogradova[1], A. V. Mihajlova[1], V. V. Paramonova[1]

(1. A.A. Baikov Institute of Metallurgy and Material Science, Russian Academy of Science, Leninskii pr. 49, Moscow, 119991 Russia; 2. NRC “Kurchatov Institute”, pl. Acalemika Kurchatova 1, Moscow, 123182 Russia)

The Experimental Investigation on the Flow Property of Aluminum Alloy AA6016 with the Fluid Media

Lihui Lang, Quanda Zhang, Sergei Alexandrov

(*School of Mechanical Engineering and Automation, Beihang University, Beijing 100191, China*)

ABSTRACT: In this paper, the hydraulic bulging experiments were respectively carried out using AA6016-T4 aluminum alloy and AA6016-O aluminum alloy, and the deformation properties of aluminum alloy under the conditions of thermal and hydraulic were analyzed. The aluminum alloy AA6016 was dealt with two kinds of heat treatment systems such as solid solution heat treatment adding natural ageing and full annealing, then the aluminum alloy such as AA6016-T4 and AA6016-O were obtained. In the same working environment, the two kinds of materials were used in the process of hydraulic bulging experiments. At the end of the experiments, according to the sizes of grid circles which have deformed and the wall thicknesses distribution near the fracture region, the flow properties of the materials were analyzed in detail from the angles of qualitative analysis and quantitative analysis.

KEY WORDS: aluminum alloy 6016; hydroforming technology; flow property; qualitative analysis; quantitative analysis

The Molybdenum Alloys for Welded Vacuum-tight Glass-to-metal Parts of Electronic Vacuum Devices

Burkhanov G. S., Kirillova V. M., Minina N. A., Kadyrbayev A. R., Mikhailov B. P., Sdobyrev V. V., Dement'ev V. A., Dormidontov N.A., Korenovskiy N. L., Sevostyanov M. A., Baikin A. S.

(*Baykov's Institute of Metallurgy and Materials Science of Russian Academy of Sciences, Russia*)

ABSTRACT: The article presents the results of the development of molybdenum alloys characterized by a close to the glass coefficient of linear expansion and that are suitable for the manufacture of vacuum-tight glass-to-metal parts by the method of low-temperature compression welding. Method of preparation of molybdenum alloys doped with rhenium, tantalum and vanadium, their structure and some properties are described. By means of dilatometric method it was found that low-alloy molybdenum alloys with rhenium (3 wt %) and vanadium (4 wt %) are the most suitable for perform of task to be achieved.

Elastic Mechanisms for a C17200 Alloy

PENG Lijun, HUANG Guojie, XIE Haofeng, FENG Xue, YANG Zhen

(State Key Laboratory for Fabrication and Processing of Nonferrous Metals, General Research Institute for Nonferrous Metals, Beijing 100088, China)

ABSTRACT: In this paper, the microstructure evolution of C17200 alloy was studied in the precipitation process by Transmission Electron Microscope. The tensile strength and elastic modulus of this alloy after different aging process were tested. The valence electron structure and cohesive energy of C17200 alloy at solid solution and aging process were calculated on the empirical electron theory of solid and molecules. The results show that precipitation sequence was α supersaturate solid solution→ G.P Zone→ (γ″)→ γ′→ γ with aging temperature 320℃. The relationship of tensile strength and aging time assume single-peaked curve principle. The elastic modulus turns to be stable before increasing along with the extending of time. High strength is ascribed to cutting mechanism between precipitate and dislocation in the alloy. The bonding force between solute and solvent atoms is higher than the force between the solute atoms after aging, which can improve the elastic modulus obviously.

KEY WORDS: C17200 alloy; aging; elastic modulus; empirical electron theory; valence theory; cohesive energy

Design and Investigation of III-V Solar Cells on Silicon Substrate with YSZ Anti-reflective Coating

Alexander Buzynin[1], Yuri Buzynin[2,3], Vladimir Shengurov[3], Nickolai Baidus[3], Nuofu Chen[4]

(1. Prokhorov General Physics Institute, Russian Academy of Sciences, Moscow, 119991 Russia; 2. Institute for Physics of Microstructures, Russian Academy of Sciences, Nizhny Novgorod, Russia; 3. Lobachevski State University, Nizhny Novgorod, 603950 Russia; 4. State Key Laboratory of Alternate Electrical Power System with Renewable Energy Sources, North China Electric Power University, Beijing 102206, China)

Effect of Heat Treatment on the Bending Behavior of Cu/Al Composite Wires

Zhen Yang, Haofeng Xie, Xujun Mi, Guojie Huang, Lijun Peng, Xue Feng, Xiangqian Yin
(*State Key Laboratory of Nonferrous Metals & Processes, General Research Institute for Nonferrous Metals, Beijing 100088, China*)

ABSTRACT: The Cu/Al composite wires have been widely used in cable industry. Especially in the development of aerospace, the light weight character of wires is particularly important. The wires are fabricated by a drawing process and heat treatment. Mechanical and electrical properties are measured and analyzed. The bending behavior is studied by scanning electron microscope.
KEY WORDS: Cu/Al composite wires; fracture; bending; intermetallic compounds; properties

Evolution of the Phase Morphology of Disperse Systems Consisting of Particles of Clay Minerals under the Action of MW Radiation

Chetverikova A.G., Filyak M.M., Kanygina O.N., Berdinsky V.L., Savinkova E.S.
(*Orenburg State University, Orenburg, Russia*)

ABSTRACT: The effect of MW radiation on a disperse system consisting of particles of natural clay has been investigated. The presence of the system's structural responses such as phase and polymorphic transformations has been established. By the methods of colorimetric gradation, fractal and wavelet analyzes the agglomeration of particles, described by DLA and CCA models, has been recorded.
KEY WORDS: phase morphology; disperse system; agglomeration; montmorillonite clay; microwave radiation; structural response

Discussion on Deformation of High Strength Aluminum Alloy by Mesoscopic Mechanics

ZHU Qifang[1], SUN Zeming[1], Fan Zhigang[1], Li Pu[1], Xia Wen[1], Liu Shufeng[1], SHARKEEV Yurii[2], Anna Eroshenko[2], KOMAROVA Ekaterina[2]

(1. General Research Institute for Non-ferrous Metals, 2 Xin Wai St., Beijing 100088, P. R. China; 2. Institute of Strength Physics and Materials Science of Siberian Branch of Russian Academy of Sciences, 2/4 Academicheskii pr., Tomsk 634055, Russian Federation)

ABSTRACT: In this paper, a new method mesoscopic been studied for metal material deformation to fracture process. In the polycrystalline material plastic flow process, mesoscopic structural features is obtained for translation - vortex by image correlation principle. A preliminary analytical method of material deformation is established. The behavior of plastic deformation of high strength aluminum alloy was analyzed. Grain groups deformation and amount of rotation were analyzed. The Rotation of grain groups is the reason of the formation of microcracks.
KEY WORDS: mesoscopic; fracture; deformation; rotation field grain groups

The Parallel Algorithm of Multi-billion Atom Samples Generation for Molecular Dynamics Simulations

A.M. Chuprunov, F.A. Sapozhnikov

(Russian Federal Nuclear Center– Zababakhin All-Russia Research Institute of Technical Physics, P.O. Box 245, Snezhinsk, Chelyabinsk Region, 456770, Russia)

ABSTRACT: The paper presents an algorithm and also results of parallel construction of a polycrystalline sample for molecular dynamics simulations. The simulated sample can comprise several region-filling materials and can have sizes from several nanometers up to tenths of a millimeter. Sample regions are presented as outer surfaces (boundary representation) or as a combination of solids (constructive solid geometry). Each region has the required number of specified monocrystalline grains wherein atoms are arranged. Atomic arrangement is done by translation of the elementary lattice cell with atoms in three directions up to the grain boundaries governed by the Voronoy partition. Depending on the region sizes, either a monoprocessor computer or a multiprocessor high-performance computer can be used for sample construction.

In the course of construction, the sample geometry is partitioned into zones and the number of these zones equals the number of computational processes. Each process is responsible for its spatial region and arranges atoms only in this particular region being a rectangular parallelepiped. For this purpose, the use is made of the function that identifies an atom to belong to a certain sample region. The open geometric kernel OpenCascade is used to identify if an atom belongs to a region with the boundary representation. The rate of this identification for one point is a critical parameter of an operating algorithm. Presented estimates demonstrate that the construction of multi-billion samples can take up to several days. The massage passing interface (MPI) is used for parallel construction and storage of the atomic data. When the multiprocessor computer is used for sample construction, the data on all identified atoms are in the distributed memory. The procedure of the simulated sample write from the distributed memory to the hard drive arrays takes into account limitations the molecular dynamics method itself imposes on spatial atomic arrangement.

The further stage will include developing the parallel algorithm of defect embedding in a constructed sample (dislocations, interstitial atoms, vacancies, etc.).
KEY WORDS: CPU; MPI; OpenCascade; geometry; boundary representation; molecular dynamics

Preparation of Nanometer Yttrium Oxide by Precipitation Method

Li Xingying[1,2], Wang Lei[3], Liu Zhiqiang[1,2], Jin Mingya[1,2]

(*1. Guangzhou Research Institute of Non-ferrous Metals, Guangzhou 510650, China; 2. Guangdong Province Key Laboratory of Rare Earth Development and Application, Guangzhou 510651, China; 3. Guangdong Material and Processing Research Institute, Guangzhou 510650, China*)

ABSTRACT: In this paper, we introduce the theoretical basis and technological process of preparing nanometer yttrium oxide by precipitation method. Method for controlling size, appearance and uniformity of particles in the process of preparing nanometer yttrium oxide by precipitation method was studied, the formation mechanism of agglomeration of particles and the way to eliminate it are studied as well. The production experiments results show that by using different methods in each process to control the formation of powder agglomeration, finally, a single particle size <50nm, aggregate size of D50<150nm, specific surface area of >45m^2/g, the purity of 99.99% spherical name-powder can be prepared.

KEY WORDS: nanometer yttrium oxide; powder; precipitation; agglomeration

Deoxidation Equilibrium in Iron-nickel Melts

V.Ya. Dashevskii, A.A. Aleksandrov, A.G. Kanevskii, L.I. Leont'ev

(*Baikov Institute of Metallurgy and Materials Science, Russian Academy of Sciences, Russia*)

Comparison of Ag and Zr with Same Atomic Ratio in Cu-Cr Alloy

Feng Xue, Xie Haofeng, Li Zongwu, Mi Xujun, Huang Guojie, Peng Lijun, Yang Zhen, Yin Xiangqian

(State Key Laboratory of Nonferrous Metals and Process, General Research Institute for Nonferrous Metals, Beijing 100088, China)

ABSTRACT: In order to compare the effect of Ag and Zr with same atomic ratio on mechanical properties and electrical conductivity improvement of Cu-Cr system alloys, three alloys (CuCrZr, CuCrAg and CuCrZrAg) were prepared. And the mechanical properties and electrical conductivity during different aging process had also been studied. The results showed that the tensile strength of CuCrAg and CuCrZrAg were similar at various aging temperature better than CuCrZr. While the electrical conductivity of CuCrZr was better than others.

"Vikhr" and Another DPF Devices for Fuson Application Experiment

E.V. Demina[1], V.A. Gribkov[1,2], A.S. Demin[1], S.A. Maslyaev[1], E.V. Morozov[1], M. Paduch[2], V.N. Pimenov[1], M.D. Prusakova[1], E. Zielinska[2]

(1. A.A. Baikov Institute of Metallurgy and Material Sciences (IMET), Russ. Ac. Sci., Moscow, RF;
2. Institute of Plasma Physics and Laser Microfusion (IPPLM), Warsaw, Poland)

ABSTRACT: Materials' damageability produced by short pulses of hot plasma (HPS) and fast ion streams (FIS) generated by a number of DPF devices are compared for various irradiation conditions related to Nuclear Fusion Reactors with Inertial Plasma Confinement. In these works the PF-6 and PF-1000U facilities (IPPLM, 6 kJ and 600 kJ) as well as a new DPF device Vikhr' (IMET RAS, 5.6 kJ) were used in parallel experiments in various regimes of their operation.

KEY WORDS: radiation tests; dense plasma focus; bulk and surface damage

Green Evaluation of Microwave Heating Furnace Using Fuzzy Decision-making Method and Life Cycle Assessment Methods

Xiaoying Zheng[1], Jin Chen[2], Guo Chen[3, 4], Jinhui Peng[2, 4], Rongsheng Ruan[4]

(1. Kunming Metallurgy College, Kunming 650093, P.R. China; 2. Key Laboratory of Unconventional Metallurgy, Ministry of Education, Kunming University of Science and Technology, Kunming 650093, P.R. China; 3. State Key Laboratory of Vanadium and Titanium Resources Comprehensive Utilization, Panzhihua 617000, P.R. China; 4. Key Laboratory of Resource Clean Conversion in Ethnic Regions of Education Department of Yunnan, Joint Research Centre for International Cross-border Ethnic Regions Biomass Clean Utilization in Yunnan, Yunnan Minzu University, Kunming 650500, P.R. China)

ABSTRACT: In this paper, a greenness evaluation index and methods of life cycle assessment of microwave heating furnace were established. Greenness evaluation index, evaluation system and evaluation model of microwave heating furnace were established. Also, life cycle assessment processing based on these ideas was introduced. Analytic hierarchy process and fuzzy decision-making method were utilized to establish and solve the greenness evaluation of assessment model, while the scores of green degree in microwave heating furnace were obtained. The results show that life cycle assessment can be applied effectively and efficiently to the microwave heating furnace.

KEY WORDS: microwave heating furnace; life cycle assessment; analytic hierarchy process; fuzzy decision-making method

The Problems of Obtaining Multilayer Nanowires Ni/Cu

I.M.Doludenko[1], D.L. Zagorskiy[2], S.A. Bedin[2,3], G.G.Bondarenko[1], V.V.Artemov[2]

(1. National Research University Higher School of Economics, Moscow,Myasnitskaya Ulitsa 20; 2. Centre of Crystallography and Photonics RAS, Moscow, Leninskii pr. 59; 3. Moscow Pedagogical State University, Moscow, Malaya Pirogovskayast 1/1)

ABSTRACT: The present work is devoted to obtaining multilayer nanowires by electrochemical filling of polymer matrices obtained by track technology. In this work, the optimal regimes of electrodeposition of nickel and copper were determined experimentally-based on the deposition of continuous layers from two-component solutions. On the basis of these data, thin layers of variable composition were obtained, and then-heterostructured nanowires (NW) consisting of alternating layers of Ni / Cu multilayers. Investigation of the structure and composition of the deposited layers and NW was carried out using the method of scanning electron microscopy and point element analysis.

Preparation and Characterization of Single Crystal Type $LiNi_{0.8}Co_{0.1}Mn_{0.1}O_2$ as Cathode Material for Lithium-ion Batteries

Xuequan Zhang[1,2], Minchao Ru[1,2], Yanbin Chen[1,2]

(1. Beijing General Research Institute of Mining & Metallurgy, Beijing 100160, China; 2. Beijing Easpring Material Technology Co., Ltd., Beijing 100160, China)

ABSTRACT: Single crystal type $LiNi_{0.8}Co_{0.1}Mn_{0.1}O_2$ (NCM811) has been synthesized from small size spherical precursor. A hexagonally ordered, layered structure with a I(003)/I(104) ratio of 1.35 was obtained. The single crystal NCM811 shows an initial discharge capacity of 196.5 mAh/g. More than 96% of the capacity is retained after 100 cycles at the 1.0 C rate in a cut-off voltage range of 4.3~3.0 V, showing better cycling stability than the normal spherical agglomerated counterpart.

Influence of the Millisecond Laser Irradiation on the Morphology and Crystallization of a Thermo-activation Zone in Bulk Amorphous Alloy Zr-Cu-Ag-Al

Fedorov V.A., Shlykova A.A., Gasanov M.F.

(ТГУ имени Г.Р. Державина, Россия)

Effect of Fe_2O_3 Content on Thermal Insulation Properties of YSZ Thermal Barrier Coatings

Ji Xiaojuan[1,2,3], Yu Yueguang[2,3], Wei Yudao[2,3], Li Yujie[2,3], Gao Lihua[2,3]

(*1. Northeastern University, School of Materials Science and Engineering, Shenyang 110819, China; 2. Beijing General Research Institute of Mining and Metallurgy, Beijing 100160, China; 3. Beijing Key Laboratory of Special Coating Materials and Technology, Beijing 102206, China*)

ABSTRACT: Compared with YSZ, Fe_2O_3 is a low melting point oxide impurity, which exists in YSZ thermal barrier coating, and has different effects on its performance. Heat insulation is an important application performance of thermal barrier coatings. The paper designed 11 kinds of YSZ coatings containing different contents of Fe_2O_3, which were prepared by APS process. The thermal diffusivity of the coatings were measured by laser method under the same conditions. The thermal expansion property was measured too. The morphology of coatings were compared using SEM. The results show that with the increase of Fe_2O_3 content in the coating, the thermal diffusivity of the coating increases at the same temperature point, that is to say, the thermal insulation ability of the coating decreases. Correspondingly, the porosity of the coating decreases and the thermal expansion contraction at high temperature is intensified, and the sintering resistance of the coating is reduced.

KEY WORDS: thermal barrier coating; Fe_2O_3; porosity; thermal expansion; thermal diffusivity

Biologically Active Agents Based on Colloidal Selenium

Folmanis G.E., Kovalenko L.V., Fedotov M.A., Volchenkova V.A.

(*A.A. Baykov's Institute of Metallurgy and Material Science Russian Academy of Sciences, Russian Federation*)

ABSTRACT: The laser ablation is a promising technology of preparation of bioactive colloidal solutions to produce high-efficient nanosized agents for different applications. Based on the colloidal selenium solution, the preparation is developed, which has the properties stimulating the plant growth, can be accumulated in the plant tissues, and protect them from diseases. The secondary laser irradiation of the formed particles results in their fragmentation, which changes their initial size distribution and a great amount of nanoparticles appears. The treatment of plants by the preparation based on the nanosized selenium allows one to solve simultaneously several problems, namely: to enrich plants with the necessary microelement (selenium), i.e. with the substance that is important for the growth and development of plants; to protect these plants from diseases; and to limit the use of harmful chemically synthesized substances in practice.

Microstructure and Residual Stress Distribution of TA15/BTi6431S Titanium Alloys Welding Joints by Ultrasonic Impact Treatment

Xiaoyun Song, Guanglu Qian, Wenjun Ye, Songxiao Hui

(State Key Laboratory of Nonferrous Metals & Process, General Research Institute for Nonferrous Metals, Beijing 100088, China)

ABSTRACT: TA15 (Ti-6.5Al-2Zr-1Mo-1V) and BTi6431S (Ti-6.5Al-3Sn-3Zr-3Mo-3Nb-1W-0.2Si) titanium alloy plates were welded through gas tungsten arc welding (TIG). The effects of ultrasonic impact treatment (UIT) on the microstructure and distribution of residual stress for the welding joint were investigated through optical microscopy and X-ray diffraction (XRD). After TIG welding, the structure of welding joint is composed of fusion zone (FZ), heat-affected zone (HAZ) and base metal. The FZ is widmannstatten structure which consists of coarse β grains and a large number of acicular α due to the fast cooling rate. The microstructure of the HAZ near base metal is similar to the base mental, that is, the microstructure near TA15 alloy is composed of coarse equiaxed α phase and that of BTI6431S alloy is coarse lamellar α phase. The residual stress in fusion zone was mainly tensile stress and the maximum longitudinal stress locate in the centerline of welding joint. The number of impact treatment has influence on residual stress distribution. After employing UIT twice, the residual stress near the welding joint exhibited a uniform distribution and the maximum tensile stress transformed to compressive stress.

KEY WORDS: titanium alloy; ultrasonic impact treatment; welding joint; residual stress; microstructure

Impulse Activation of Hydrogen-air Fuel Cell

Elena Galitskaya[1,2], Ekaterina Gerasimova[3], Yuri Dobrovolsky[3], Alexander Sivak[1], Vitaly Sinitsyn[1,2]

(1. In Energy Group, Elektrodnaya 12-1, 111524, Moscow, Russia; 2. Institute of Solid State Physics RAS, Academician Ossipyan 2, 142432, Chernogolovka, Moscow distr., Russia; 3. Institute of Problems of Chemical Physics, Russian Academy of Sciences, 142432, Chernogolovka, Moscow distr., Russia)

Quasi-static and Dynamic Properties of Ti-4Al-3V-0.6Fe-0.2O Titanium Alloy Plates

Rui Liu, Songxiao Hui, Wenjun Ye, Yang Yu, Xiaoyun Song

(State Key Lab of Nonferrous Metals & Processes, General Research Institute for Nonferrous Metals, Beijing 100088, China)

ABSTRACT: This paper aimed to study the quasi-static and dynamic properties of the titanium alloy with nominal composition Ti-4Al-3V-0.6Fe-0.2O. The ingots Ti-4Al-3V-0.6Fe-0.2O alloy were forged into slabs of 32 mm thickness by one heat above $T_{\beta\text{-transus}}$ temperature and one heat below $T_{\beta\text{-transus}}$ temperature, and then rolled into plates of 8 mm thickness by one heat below $T_{\beta\text{-transus}}$ temperature. The quasi-static tensile properties were tested by MTS™ testing system at strain rate of 10^{-3} s^{-1}. The dynamic compression properties were tested by Split Hopkinson Pressure Bar system at strain rate of (3000±200) s^{-1}. The results show that the quasi-static properties of the Ti-4Al-3V-0.6Fe-0.2O plate are comparable to ATI 425™ alloy plate and commercial Ti-6Al-4V plate. The average dynamic flow stress of Ti-4Al-3V-0.6Fe-0.2O plate is comparable to ATI 425™ alloy plate, over 100MPa higher than that of Ti-6Al-4V plate. The maximum strain during homogeneous dynamic plastic deformation of Ti-4Al-3V-0.6Fe-0.2O plate is approximately 15%-20% higher than that of ATI 425™ alloy plate, and is approximately 80% of that of Ti-6Al-4V plate.

KEY WORDS: titanium alloy; mechanical properties; dynamic properties

Layer Laser Fusing of Sm and Fe Powders on the Non-magnetic Stainless Steel Substrates with and Without Additional Magnetic Field: Composition and Ferromagnetic Properties

N.G. Galkin[1,2], Y.N. Kulchin[1,2], E.P. Subbotin[1], A.I. Nikitin[1], D. S. Yatsko[1], M.E. Stebliy[2], A.V. Nepomnyaschiy[1]

(*1. Institute of Automation and Control Processes of Far Eastern Branch of Russian Academy of Sciences, Radio Str., 5, 690041, Vladivostok, Russia; 2. Far Eastern Federal University, Vladivostok, 690950, Sukhanova Str., 8, Russia*)

ABSTRACT: The layer laser fusing (LLF) of Sm powder and then the Fe powder were carried out on the robotic laser technological complex (RLTC) for the creation of (0.5-1.5) mm thickness ferromagnetic coatings on the stainless steel substrate in argon flow conditions. It was established that after LLF process the oxidation of Sm and Fe occurs in the subsurface region, but inside the formed coating the oxygen concentration was closed to zero and the formation of two types of Fe-Cr-Sm and Fe-Sm alloys with different compositions and electron work function were confirmed. Two types of laser fusing were tested. At the additional magnetic field (0.2 T) the crystallization of single grains (5-20 μm) of Fe-Cr-Ni alloy, with mutually perpendicular orientations and formation a Sm-Fe mesh structure in the mentioned alloy matrix was found across a thickness of the fused layer (1.0-1.3 mm). When the additional magnetic field is absent during LLF process two regions on the depth of laser fused coating were observed. Near the surface the Sm oxide layer was formed with limited thickness (smaller 10 μm). At larger thickness the formation of high density of elongated (down to depth of 100-600 μm) and oval form (2-10 μm in diameter) Sm-Fe-Cr-Ni alloy's grains without oxygen concentration was observed. The layer laser fusing of Sm and Fe powders has resulted to formation of coatings with soft ferromagnetic properties at 300 K with low coercivity (20 - 100 Oe), near zero residual magnetization and high saturation magnetization (110-112 emu/g) that is 55% from the same value of the original Fe powder. It was established that additional magnetic field only influenced on the coercivity due to the mutually perpendicular orientation of ferromagnetic Fe-Cr-Sm grains. The cooling of both Fe-Sm coatings has shown a magnetic ordering with 50 K Curie temperature.

KEY WORDS: laser fusing; samarium; iron; powders; magnetic properties; microstructure

The Development Trend of Zinc Electrodeposition Technology

Xie Gang, Hu Fengying, Ma Jilun

(*China Nonferrous Metals Industry Technical Development and Exchange Center, Beijing 100814, China*)

ABSTRACT: The electrochemical system of zinc electrodeposition and the production process of electrolysis zinc from zinc sulfate solution are introduced in this paper. It also described the influence of different factors on the production process of zinc electrodeposition. By comparing the size of plate and the way of stripping zinc, the advantages of the large plate and the automatic zinc stripping technology are obtained.

KEY WORDS: zinc electrodeposition; large plate; automatic zinc-stripping

Investigations of Young's Modulus and Microhardness of Laser-fused Sm-Fe and Sm-Co Coatings on Two Types of Substrates

N.G. Galkin, Y.N. Kulchin, D. S. Yatsko, E.P. Subbotin, A.I. Nikitin, A.V. Nepomnyaschiy

(*Institute of Automation and Control Processes of Far Eastern Branch of Russian Academy of Sciences, Radio Str., 5, 690041, Vladivostok, Russia*)

ABSTRACT: Studies on Young's modulus and hardness of laser-fused surfaces: Sm-Co on the stainless steel and Sm-Fe on the duralumin substrates have shown the heterogeneity of depth properties with local maximum in areas not containing microcracks. Maximum hardening was obtained for Sm-Fe system. Found that hardening of laser-fused layers and the increase their resistance to compression and tension on the stainless steel substrate is associated with the formation of two and three-component alloys ($Fe_{17.8}Ni_{25.6}Cr_{5.6}C_{11.0}Sm_{32.1}$and $Fe_{64.3}Ni_{1.9}Cr_{16.6}C_{5.5}Sm_{8.4}$) due to the diffusion of atoms of the substrate and depends on the availability of the microcrack's grid of and the alloy microstructure. The high Al diffusion speed from duralumin substrate in the fused layer, formed from Sm powder, and then from Fe powder, led to the segregation of Al on the surface, formation of triple oxide ($AlSmO_3$) with maximum hardness and Young's modulus. Further decrease of both parameters is associated with the formation of microcracks, but a local maximum connects with strong decreasing of microcrack's density and formation of Al_2Sm layer.

KEY WORDS: laser fusing; samarium; iron; powders; magnetic properties; microstructure

Investigation on High Performance (γ+α_2+B2) Multi-phase TiAl Alloy

Zhenxi Li, Hongwu Liu, Fan Gao, Chunxiao Cao

(*AECC Beijing Institute of Aeronautical Materials*)

ABSTRACT: With the aim of developing a new TiAl alloy with excellent mechanical properties, a novel multi-phase Ti-(42-44)Al-6(Cr, V) alloy (G9 alloy) was investigated systematically. The hot deformation behavior and microstructural evolution of as-cast and as-extrusion annealed G9 alloy were studied by hot compression tests. It was indicated that the optimum hot working condition of as-cast G9 alloy occurred in the temperature range of 1175-1225℃ and the strain rate range 0.05-0.1s^{-1}. The large-size TiAl alloy rectangular bars with crack-free appearance were successfully prepared by hot extrusion. After annealing, the fine and uniform microstructure with excellent deformation ability was obtained, the average grain size was 13.7μm. Using hot pack rolling technology, the large-size (300mm×200mm×2.7mm) G9 sheets were successfully fabricated. The microstructure of the sheet was mainly composed of equiaxed γ grains, irregular B2 grains, and little massive α_2. Its average grain size was 3.5μm. At 950℃/$1\times10^{-4}s^{-1}$, the superplastic elongation reached 345%. At room temperature, the tensile strength of G9 alloy was 1113 MPa and the tensile ductility was 2.4%, demonstrating a good combination of high strength and ductility. Even up to 800℃, the as-extruded G9 alloy maintained a tensile strength more than 860MPa. TiAl superplastic forming parts with the application background were successfully prepared, which were of great significance to promote the application of TiAl alloy sheet.

KEY WORDS: multi-phase TiAl alloys; mechanical properties; microstructure; B2 phase

Localized Dissolution of Carbon Steel under Cathodic Polarization Conditions

Natalia Gladkikh, Marina Maleeva, Maxim Petrunin, Ludmila Maksaeva
(*Frumkin Institute of Physical Chemistry and Electrochemistry RAS, Russian*)

ABSTRACT: The most widespread method of protection of underground constructions against corrosion is electrochemical protection (ECP) which effectively reduces metal corrosion speed. However, it is known that one of the most dangerous types of corrosion destruction – stress corrosion cracking (SCC) appears on underground pipelines at which the ECP works. It specifies that at cathodic potential local corrosion processes can proceed. As emergence of defects often is the first stage of corrosion cracking, the purpose of the real work was studying of processes of emergence and development of local (looks like pitting) defects at cathodic polarization and identification of the factors defining intensity of processes of defect formation.

Observation of change of a surface of a working sample was carried out in borax buffer solution (pH6.7) with sodium chloride additives with use of "in situ" of optical microscopy, shooting the image of a surface with the video camera the Received images processed in the graphic editor. Electrochemical tests carried out by means of a potentiostat in a three-electrode cell. Simultaneous use of optical and electrochemical methods of research has allowed recording emergence of defects in the field of cathodic protection of underground constructions. The subsequent processing of images has allowed studying in detail kinetics of initial stages of local dissolution of steel at cathodic potentials. Influence of size of potential on emergence and development of defects has been investigated. With high precision are defined: the incubator period of appearance of defects, speed of filling of a surface of metal with defects, kinetics of development of individual defects on a surface.

As potential shift to the area of negative values is followed by increase in speed of a hydrogen-charging of metal, investigated influence of penetration of electrolytic hydrogen into metal on processes of emergence of local defects for what added to working electrolyte a promoter of a hydrogen-charging thiurea, in parallel measuring the speed of penetration of hydrogen into steel. It is established that even at small concentration to 1 mm thiurea promotes an intensification of process of defect formation at cathodic potentials.

KEY WORDS: local cathodic defects; corrosion image

Corrosion Resistance of UV Curable Hybrid Coatings on the Galvanized Sheet

Fengguo Liu[1], Wenju Tao[1], Yuzheng Wang[2], Youjian Yang[1], Xianwei Hu[1], Zhaowen Wang[1], Xiangxin Xue[1]

(*1. School of Metallurgy, Northeastern University, Shenyang 110819, China*; *2. School of Materials Science and Engineering, Shenyang University of Technology, Shenyang 110870, China*)

ABSTRACT: UV curable hybrid coating is a green environment-friendly coating with the advantages of zero VOC emission, energy saving and high curing rate. In this work, UV curable hybrid coatings based on the epoxy acrylate were prepared. The hybrid coatings could be applied onto the pretreated galvanized sheets. The adhesion could reach the requirements of national standards. The corrosion resistances in NaCl solution were investigated for three types of coatings by electrochemical impedance spectroscopy: the pure EA coating, EA-SiO_2 coating and EA-Si coating. As a summary the corrosion resistance of the three coatings on metal substrates prepared in this study was found to be in the order of EA-Si coating > EA-SiO_2 coating > pure EA coating.

KEY WORDS: hybrid coatings; UV curing; adhesion; EIS; corrosion resistance

Surface Modification of Aluminum in the Plasma of a Hollow-cathode Non-self-sustained Glow Discharge

Ivanov Yu. F., Tolkachev O. S., Denisov V. V., Petrikova E. A., Krysina O. V.

(*Institute of High Current Electronics, SB, RAS, Russia*)

Characteristics of AA4045/AA3003 Clad Hollow Billet Prepared by Direct-Chill Process

Xing Han[1], Shuncheng Wang[1], Deng Nong[1], Kaihong Zheng[1], Haitao Zhang[2], Jianzhong Cui[2]

(1. Guangdong Institute of Materials and Processing, No. 363 Changxing Road, Tianhe District, Guangzhou, 510651 P. R. China; 2. Key Laboratory of Electromagnetic Processing of Materials, Ministry of Education, Northeastern University, No. 3-11, Wenhua Road, Heping District, Shenyang 110819, P. R. China)

ABSTRACT: AA4045/AA3003 clad hollow billet with size of ϕ140mm/ϕ110mm/ϕ70mm was prepared by direct chill casting process. The macrostructures, microstructures, compositions distribution, micro-hardness and the mechanical properties of the interface were investigated in detail. The results show that the cladding billet with few defects could be obtained by direct chill casting process. At the interface, diffusion layer about 10μm on average formed between the two alloys due to the diffusion of alloys elements. From AA4045 side to AA3003 side, Si content has a trend to decrease, while Mn content has a trend to increase gradually. Tensile strength of the cladding billet reaches 112.7 MPa and the shearing strength is 75.6 MPa, revealing that the two alloys were combined metallurgically by mutual diffusion of alloy elements.

KEY WORDS: clad hollow billet; direct chill casting; diffusion layer; metallurgical bonding

Core-shell and Bi-phasic Structure Magnetic Nanoparticles for Biomedical Applications

Kamzin A.S.[1], Phan M. H.[2], Tkachenko M. V.[3]

(1. Ioffe Physical-Technical Institute RAS, S-Petersburg, Russia;
2. Department of Physics, University of South Florida, Tampa, Florida 33620, USA;
3. Karazin Kharkiv National University, Sq. Svobody 4, Kharkiv, 61022 Ukraine)

Preparation of TiO_2 Microfiltration Membrane on Porous Stainless Steel Tube

Shuai Li, Chao Zhang, Qinli Lv, Hua Zhang, Di He, Shumao Wang, Lijun Jiang

(*General Research Institute for Nonferrous Metals, Beijing, China*)

ABSTRACT: TiO_2 microfiltration composite membrane was prepared on porous stainless steel tube by using a dip-coating method. An intermediate layer made of porous stainless steel was introduced between the substrate and TiO_2 membrane. The forming process of the composite membrane was analyzed by TG-DSC and XRD. The microstructure of the membrane was characterized by SEM. Bubble point method was used to analyze the pore size distribution of the microfiltration membrane. Homogeneous TiO_2 membrane with good bonding strength was obtained on the stainless steel tube. The TiO_2 membrane has an average pore size of 0.09μm.

Микроструктурные Аспекты Хладноломкости Низколегированных Сталей

Кантор М.М., Воркачев К.Г., Солнцев К.А.

(*Институт металлургии и материаловедения им. А.А. Байкова РАН, Россия*)

An Efficient Method for the Synthesis of SiBNC Preceramic Polymers with Different Si/B Atomic Ratios

Ji Xiaoyu, Shao Changwei, Wang Hao

(Scienceand Technology on Advanced Ceramic Fibers and Composites Laboratory, National University of Defense Technology, Changsha 410073, P.R. China)

ABSTRACT: Random inorganic networks composed of silicon, boron, nitrogen and carbon represent amorphous SiBNC ceramics with outstanding durability up to 1800℃. The high-temperature durability of SiBNC ceramics is significantly influenced by Si/B ratios and the synthetic procedures. This work was aimed to study the Si/B atoms ratio of the PBSZ and the high temperature properties of the pyrolyzed products. Three kinds of PBSZ polymers were obtained via ammonolysis reaction of the trichloroborane and dichloromethylsilane with hexamethyldisilazane and the mole ratios are 1 : 0.5 : 6, 1 : 1 : 6 and 1 : 2 : 6, respectively. The prepared polymers and their derived ceramics were investigated by elemental analysis and spectroscopy techniques including FTIR, MAS NMR, XPS, SEM, and TEM. The random arrangement of the -N-Si(CH_3)- and BN_3 units formed the desired precursors with -Si-N-B- frameworks, accompanied with a few -Si-N-Si- and -B-N-B- units. The precursors were converted into multinary ceramics when heated above 800°C under N_2 and transformed into $SiB_{2.6}N_5C_{2.2}$, $SiB_{0.9}N_{2.7}C_{1.3}$ and $Si_2BN_3C_{1.4}$ ceramics respectively at 1000°C with the yields of 43.2wt%, 50.1wt%, and 63.2wt%. Under N_2, the obtained SiBNC ceramics all remained amorphous up to 1600℃. $SiB_{0.9}N_{2.7}C_{1.3}$ was confirmed to have the best resistance against crystallization, and remained amorphous until 1700℃. The agglomeration of B-N-B linkages ($SiB_{2.6}N_5C_{2.2}$) and Si-N-Si linkages ($SiB_2N_3C_{1.4}$) damaged the polyhedral B-N-Si construction, which played an important role in the resistance against crystallization of BN and Si_3N_4. Using the precursor with equal Si/B ratio, the SiBNC ceramic fibres were also obtained.

KEY WORDS: SiBNC; Si/B; polyborosilazane; soften point

Laser-interference Control of the Strain Field in the Process of Manufacturing Details

Kesariiskyi O.G.[1], Kondrashchenko V.I.[2]

(1. Laboratory of Complex Technologies Ltd., Ukraine; 2. Russian University of Transport, Russia)

EDC-induced Self-assembly of BSA-Au NCs

Wenli Zhu[1], Huili Li[2], Ajun Wan[3]

(1. College of Chemistry and Chemical Engineering, Shanghai Jiaotong University, Shanghai 200240, China; 2. College of Pharmacy, Shanghai Jiaotong University, Shanghai 200240, China; 3. College of Medicine, Tongji University, Shanghai 200092, China)

ABSTRACT: The bovine serum albumin stabilized gold nanoclusters (BSA-Au NCs) has attracted great attraction due to BSA-Au NCs being more readily bioconjugated, more stable and suitable for high sensitivity, selectivity and rapid detection of analytes. In this paper, the small molecule N-ethyl-N′-(dimethylaminopropyl) carbodiimide (EDC) was used as the cross-linking agent to induce the self-assembly of BSA-Au NCs. In the presence of EDC, BSA-Au NCs self-assembled into wafer-shape Au NPs. As the concentration was 0.0020M, the particle size was uneven. When EDC concentration was 0.0025 and 0.0030M, the particle size of Au NPs distributed uniformly. As EDC concentration was higher than 0.0030M, the Au NPs microstructure changed from smooth and compact to rough and loose structure. The particle size of Au NPs increased with EDC concentration. When EDC concentration was lower, EDC bioconjugated with BSA, resulting in the structure instability of BSA-Au NCs. EDC-conjugated surface was more unstable than the non-reacted area. The unstable surface caused the reaction with the other Au NCs for the reconstruction of smooth and compact Au NPs based on the mechanism of Ostwald ripening. However, when EDC concentration was higher, branched Au NPs structures were destroyed by high concentrations of EDC, which hindered the ripening of Au NPs. The self-assembled Au NPs remained the unique fluorescence characteristics. When EDC concentration was higher than 0.0020M, the fluorescence intensity of Au NPs was even higher than that of BSA-Au NCs. Moreover, when EDC concentration was 0.0025M, the fluorescence intensity was the highest.

Influence of Nanosecond Laser Radiation on the Surface Structure of Metals

R.R. Khasaya, Yu.V. Khomich, S.I. Mikolutskiy

(Institute for Electrophysics and Electric Power of Russian Academy of Sciences)

Plasma Discharge Electrolysis of Nano Cuprous Oxide and Copper Particle in Solution

Lingling Shen

(Northeastern University, Institute of Metallurgy, Shenyang 110819, China)

ABSTRACT: Cu_2O and Cu micro-particle were facilely synthesized by solid-liquid plasma discharge electrolysis method in solution of cupric sulfate and sulphuric acid added respectively. The phase component, morphologies and mechanism were investigated by XRD, SEM/EDS and emission spectrum. In cupric sulfate electrolyte, a higher concentration benefit to the formation of Cu_2O compared with Cu phase, while the product phase composition depends on relative proportion of sulphuric acid and copper sulphate for Cu formation process in mixture of $CuSO_4 + H_2SO_4$. The formation mechanism of Cu_2O can be described as $2Cu^{2+} + H_2O + 2e^- = Cu_2O + 2H^+$ with a decrease of pH value in copper sulphate solution. While a near-pure Cu phase can be formed in solution of 0.2M $CuSO_4$ + 0.2M H_2SO_4, which attributed to the existence of reducing agent H· according to $Cu^{2+} + 2H\cdot = Cu + 2H^+$, and the pH value increased due to few amount of Cu^{2+} in reaction region and recombination between hydrogen radicals. Moreover, the relationship of discharge between Cu^{2+} and H^+ is competitive in $CuSO_4 + H_2SO_4$. The morphology of Cu_2O was polyhedron shape and dispersive with a diameter of 100nm, on the contrary, synthesized Cu phase had a nearly spherical shape, but aggregation growth phenomenon with a mean diameter of 200nm.

Low-temperature Curing High-silica Material with Increased Corrosion Resistance

Klimenko N.N., Levina Yu.N., Mikhailenko N.Yu.

(Dmitry Mendeleev University, Russia, klimenko)

C103 铌合金表面 Si-Cr-Ti 熔烧硅化物涂层 1400℃高温抗氧化性能研究

汪　欣，李争显，杜继红，李晴宇
(*西北有色金属研究院，西安 710016*)

摘　要： 铌基合金熔点高、高温强度好且具有良好的加工性能，是重要的高温结构材料，但其氧亲合势高，易发生氧化，制约了其在高温氧化环境中的应用。本研究采用料浆真空高温熔烧方法在 C103 铌合金表面制备了 Si-Cr-Ti 高温抗氧化涂层，测试了涂层在 1400℃静态空气中的高温抗氧化性能，利用 SEM、EDS、XRD 等仪器分析了涂层氧化前后的显微结构、元素分布和相组成。研究结果表明，涂层由复杂的$(Nb,Cr,Ti)Si_2$相组成，涂层内部无贯穿性裂纹，涂层与基体之间形成了冶金结合；涂层制备及后续氧化过程中涂层与基体之间发生了互扩散，扩散区由 $NbSi_2$ 相和 Nb_5Si_3 相组成，过渡层无孔洞，能有效阻止涂层内部裂纹的进一步扩展以及氧向基体一侧的扩散，从而提高涂层的高温抗氧化性能。氧化过程中，涂层表面生成一层玻璃膜，该玻璃膜阻止了氧向涂层内部的扩散，且该玻璃膜具备高温自愈合能力。
关键词： 料浆烧结；铌合金；硅化物涂层；高温氧化

Oxidation Behavior of Slurry Fused Si-Cr-Ti Silicide Coating on C103 Alloy at 1400℃

Xin Wang, Zhengxian Li, Jihong Du, Qingyu Li
(*Northwest Institute for Non-Ferrous Metal Research, Xi'an 710016, China*)

ABSTRACT: Because of high melting points, good high temperature mechanical properties and excellent machinability, Nb-based alloys have been proved to be of great importance as high temperature structural materials. In this research, Si-Cr-Ti silicide coating was prepared on C103 Nb-based alloy by slurry fusion, and its oxidation behavior was tested at 1400℃. Methods of SEM, EDS and XRD were employed to characterize the microstructure, elemental distribution and phase composition of the coating. The silicide coating adhered to the C103 substrate very well and was composed of $(Nb,Cr,Ti)Si_2$ without any perfoliate cracks. Interdiffusion occurred between the silicide coating and the substrate, and phases of $NbSi_2$ and Nb_5Si_3 formed in the interdiffusion zone. The pore-free interdiffusion zone was proved able to prohibit the propagation of cracks within the coating towards the substrate as well as the oxygen permeability, which was beneficial for improving the oxidation resistance. During oxidation, a layer of glass formed above the coating due to the selective oxidation of Si within the coating, and the self-healing glass scale could protect the substrate from rapid oxidation at 1400℃.
KEY WORDS: slurry fusion; Nb-based alloy; silicide coating; high temperature oxidation

Structural Designing of High-coercivity (Nd, Pr, Dy, Tb)-Fe-B Magnets via *REM*H_2 Hydride Adding

Kolchugina N.B.[1,3], Burkhanov G.S.[1], Lukin A.A.[2], Koshkid'ko Yu.S.[3], Skotnicova K.[4], Cwik J.[3], Rogacki K.[3,5], Drulis H.[5]

(1. Baikov Institute of Metallurgy and Materials Science, Russian Academy of Sciences, Leninskii pr. 49 Moscow 119334 Russia; 2. "JSC SPETSMAGNIT", Dmitrovskoe Sh. 58, Moscow, 127238 Russia; 3. International Laboratory for High Magnetic Fields and Low Temperatures, Polish Academy of Sciences, Gajowicka 95, 53-421 Wroclaw, Poland; 4. Vysoka Skola Banska - Technical University of Ostrava, 17 Listopadu, 15/2172, 70833, Ostrava-Poruba, Czech Republic; 5. Institute of Low Temperature and Structure Research, Polish Academy of Sciences, Okolna 2, 50-422 Wroclaw, Poland)

Preparation and Photocatalytic Activity of TiO_2 Powder Codopd with La and Graphene Oxide

Wei Wang, Limei Yang, Songtao Huang, Zheng Xu, Zhen Cheng, Lu Jia

(National Engineering Lab. of Biohydrometallurgy, GRINM Engineering Institute, Beijing 101407, China)

ABSTRACT: La-TiO_2 powder, graphene oxide(GO)-TiO_2 powder and TiO_2 powder codoped with La and GO were successfully synthesized by sol-gel method and characterized by XRD, UV-Vis DRS, TEM and PL. The results show that the La doping decrease the crystal size from 25nm to 15nm, prevent the combination of photogenerated electrons and holes, reduce the forbidden band width of TiO_2 and promote the red shift of the absorption spectrum. In addition, GO doping also prevent the combination of photogenerated electrons and holes to improve the quantum efficiency of TiO_2 and reduced the forbidden band width of TiO_2. It was observed that the light absorption edges of TiO_2 was moved to longer wavelengths and could effectively promote the utilization of the visible light when adding GO. Compared with photodegradation rates of methylene blue of the La-TiO_2 powder and GO-TiO_2 powder, the TiO_2 powder codoped 0.50% La and 0.50% GO had the best photocatalytic performance.
KEY WORDS: photocatalyst; TiO_2; lanthanum; graphene oxide

Phase Transformation and Morphology in the Process of Sythesis of YAG:Nd

Kolomiets T.Yu., Telnova G.B., Solntsev K.A.

(Baikov Institute of Metallurgy and Materials Science, Russian Academy of Sciences, Leninskii pr. 49, Moscow, 119991 Russia)

ABSTRACT: Using high-temperature X-ray diffraction, differential scanning calorimetry, and electron microscopy, we have studied the formation of yttrium aluminates and Nd:YAG (YAG) activated garnet nanoparticles during the thermal decomposition of a poorly crystallized carbonate precursor prepared in the $NH_4Al(OH)_2CO_3$–$(Y,Nd)(OH)CO_3$ nanosystem and the development of the morphological structure of powders during heating to a temperature of 1350℃. The results demonstrate that heat treatment in the temperature range 850–950℃ leads to the formation of metastable nonstoichiometric $YAlO_3$ with a garnet-like structure, which reacts with Al_2O_3 at a temperature of 1000℃ to form YAG. The cubic cell parameter and X-ray density of YAG crystals with the composition $Y_{2.97}Nd_{0.03}Al_5O_{12}$ synthesized at 1200℃ are 1.2009 nm and 4.565 g/cm^3, respectively, and the average particle size is 108 nm. Using carbonate route, we prepared transparent Nd:YAG ceramics with a relative density of 99.7%, X-ray density of 4.562 g/cm^3, and crystallite size in the range 1–7 μm.

Additive Manufacturing of Strong and Ductile Cu-15Ni-8Sn

Chao Chen, Gengming Zhang, Kechao Zhou

(State Key Laboratory of Powder Metallurgy, Central South University, Changsha 410083, China)

ABSTRACT: Cu-15Ni-8Sn specimens are manufactured by selective laser melting (SLM) from gas-atomized powders and their microstructures and mechanical properties have been analyzed by mean of optical microscope, electron probe microanalysis, scanning electron microscopy and tensile-test facility. The relative density of SLM samples increase with increasing power density and reach peak value of 99.4% without any post-treatment. Room-temperature tensile tests reveal a remarkable mechanical behavior. The sample of 99.4% relative density shows yield and tensile strengths of about 522MPa and 653MPa, respectively, along with fracture strain of 17%. The result is also compared to those samples fabricated by traditional powder metallurgy. The high strength and excellent ductility of Cu-15Ni-8Sn can be ascribed to the refined microstructure resulting from the high cooling rate imposed by laser processing, further demonstrating the effectiveness of SLM for the production of materials with enhanced mechanical performance.

Model of Concrete Macrostructure

Kondrashchenko V.I.[1], JING Guoqing[2], Kondrashchenko E.V.[3], WANG Chuang[1]

(1. Russian University of Transport, Russia; 2. Beijing Jiaotong University, China; 3. A. N. Beketov National University of Urban Economy in Kharkov, Ukraine)

Comparative Study on Activity of Zn-Based Multicomponent Liquid Alloys between Different Thermodynamic Models

Heng Dai, Dongping Tao

(Faculty of Metallurgical and Energy Engineering, Kunming University of Science and Technology, Kunming 650093, Yunnan, P.R. China)

ABSTRACT: The Wilson model, NRTL and the molecular interaction volume model (MIVM) are all depend on the concept of local composition, and the sub-regular solution model (SRSM) is improved from the regular solution model (RSM). In order to compare their prediction effects, the activities of Zn in the Zn-Ag-Sb, Zn-Cu-In and Zn-Cu-Sb ternary systems and in the Zn-Sn-Cd-Pb quaternary system are predicted by the MIVM, NRTL, Wilson model and SRSM, respectively. The predicted values are compared with their corresponding experimental data. The results show that average standard deviations of MIVM, Wilson, NRTL and SRSM are 0.0306, 0.0703, 0.0504 and 0.0655 respectively, and their average relative errors are 10%, 17%, 15% and 30% respectively. Therefore, the predicting ability of these four models is as sequence as MIVM>NRTL>Wilson>SRSM, and among them the local composition models are of better stability than SRSM and MIVM is more suitable for predicting activities of Zn-based multicomponent liquid alloys.

KEY WORDS: Zn-based liquid alloys; activity; MIVM

Evaluation of Workability of Machine Parts on the Basis of the Odinga Criterion

Leonid Kondratenko[1], Viktor Terekhov[2], Lyubov Mironova[3]

(1. State Science Center RF of TsNIITMash, Moscow, Russia; 2. Moscow Engineering Physics Institute "MEPhI", Moscow, Russia; 3. Moscow State University of Railway Engineering, Moscow, Russia)

Extension of Molecular Interaction Volume Model to Electrolyte Solution

Congyu Zhang, Dongping Tao

(Faculty of Metallurgical and Energy Engineering, Kunming University of Science and Technology, Kunming 650093, Yunnan, P.R. China)

ABSTRACT: The molecular interaction volume model (MIVM) for non-electrolytes has been extended to represent excess Gibbs energy of aqueous electrolyte solutions. The model is expressed as a sum of contributions of a long range term and a short range term. The long range term is represented by Pitzer-Debye-Hückel equation; the short range term is represented by modified MIVM. The new model contains two key assumptions which are local electro-neutrality and like-ion repulsion proposed in electrolyte NTRL model. With two adjustable parameters per electrolyte, the new model is applied to correlation the mean activity coefficients of several binary aqueous electrolyte systems at room temperature. The results are compared with electrolyte-NRTL model. The comparison shows that the new model can correlate the experiment data of binary electrolytes accurately and better than the electrolyte-NRTL model because the new model has clear physical basis and thermodynamic consistency.

Shock Waves in a Zinc Single Crystal

Krivosheina M.N.[1], Kobenko S.V.[2], Tuch E.V.[1], Lotkov A.I.[1], Kashin O.A.[1]

(1. Institute of Strength Physics and Materials Science SB RAS, Tomsk, 634055, Russia;
2. Nizhnevartovsk State University, Nizhnevartovsk, 628602, Russia)

HfC 陶瓷先驱体的制备及其性能研究

张丽艳，王小宙，王亦菲

(国防科技大学，航天科学与工程学院新型陶瓷纤维及其复合材料重点实验室 长沙 410073)

摘　要: HfC 陶瓷具有优异的耐超高温性能，在航空航天领域具有广阔的应用前景。本文通过先驱体转化法，以四氯化铪，乙酰丙酮，乙醇，1,4-丁二醇为原料合成了 HfC 陶瓷先驱体。采用元素分析、红外光谱、核磁共振、TGA 等对先驱体的组成、结构及性能进行了表征。结果表明：先驱体主要含有 Hf、C、O、H 元素，先驱体是一种主链含 Hf-O 键和-C_4H_6-键，侧链含乙酰丙酮的线性聚合物，在丙酮、甲醇等溶剂中具有良好的溶解性。先驱体 1000℃陶瓷产率为 53wt%。采用元素分析、红外光谱、XRD、SEM 等对先驱体的无机化过程及陶瓷产物的组成、结构与性能进行了表征，结果表明：1000℃时陶瓷产物主要为 HfO_2，1400℃逐渐转化为 HfC 陶瓷。所制备的 HfC 陶瓷先驱体具有制备工艺简单、溶解性能优良、陶瓷产物性能好的优点，在聚合物浸渍裂解工艺制备超高温陶瓷基复合材料上有很好的应用前景。

关键词: HfC；超高温；先驱体；陶瓷

Preparation and Properties of HfC Ceramic Precursor

Liyan Zhang, Xiaozhou Wang, Yifei Wang

(Science and Technology on Advanced Ceramic Fibers and Composites Laboratory, National University of Defense Technology, 410073, Changsha, P.R.China)

ABSTRACT: As an important ultra-high temperature ceramics (UHTCs), HfC ceramics have been considered to be one of the most promising materials for the application in aerospace. A precursor for HfC ceramic was prepared by using: Hafnium(IV) chloride, methanol, acetylacetone and 1,4-Butylene glycol as raw materials The composition, structure and properties of the obtained precursor was investigated by elemental analysis, Fourier transform infrared (FTIR), nuclear magnetic resonance (NMR) and TGA. The results show that, the precursor mainly contains Hf、C、O、H, with a linear structure of Hf-C and -C_4H_6- in main chain, acetylacetone as ligand in side chain, which exhibits good solubility in tetrahydrofuran, xylene, etc. The ceramic yield of the precursor is about 53wt% at 1000℃. The composition, structure and pyrolysis of the precursor were analyzed by elemental analysis, FTIR、XRD、and SEM. The results indicate the pyrolysis product of the precursor is HfO_2 at 1000℃, which starts to translate into HfC at 1400℃. Owe to the simple synthesis, good solubility and outstanding properties of the ceramic product, HfC ceramic precursor holds great promise for the fabrication of ultra-high temperatures ceramic matrix composites via precursor infiltration and prolysis.

KEY WORDS: HfC; ultra-high-temperature; precursor; ceramic

Phase Formation and Ionic Conductivity of Zirconia-based Crystals Grown by Skull Melting Technique

A.V. Kulebyakin[1], M.A. Borik[1], S.I. Bredikhin[2], V.T. Bublik[3], I.E. Kuritsyna[2], E.E. Lomonova[1], F.O. Milovich[3], V.A. Myzina[1], V.V. Osiko[1], P.A. Ryabochkina[4], N.Yu. Tabachkova[3]

(*1. Prokhorov General Physics Institute RAS, Russia; 2. Institute of Solid State Physics RAS, Russia; 3. National University of Science and Technology 《MISIS》, Russia; 4. Ogarev Mordovia State University, Russia*)

Application of Molecular Interaction Volume Model to Fe-Based Liquid Alloys

Yan Zhao, Heng Dai, Dongping Tao

(*Faculty of Metallurgical and Energy Engineering, Kunming University of Science and Technology, Kunming 650093, Yunnan, P.R. China*)

ABSTRACT: The knowledge of thermodynamic properties of alloys is important for providing thermodynamic information in metallurgical processes. But the experimental thermodynamic study is very time consuming since a great number of measurements are necessary. Therefore, theoretical prediction is a significant and effective approach to obtain thermodynamic properties of liquid alloys, especially for multicomponent ones. Based on the molecular interaction volume model (MIVM) and its pseudo-multicomponent approach, the activities or activity coefficients of components and their phase equilibria in Fe-based liquid alloy have been calculated. For example, the component activities of Cr in liquid Fe-C-Cr and those of Fe, Cr and P in liquid Fe-Cr-P have been predicted. The results indicated that the average relative errors between predicted values and experimental data of C in liquid Fe-C-Cr was 40% and those of Fe, Cr and P in Fe-Cr-P were 6%, 16% and 22%, respectively. A significant advantage of the model lies in its ability to predict thermodynamic properties of liquid alloys using only binary infinite activity coefficients.

Electrophysical Properties of the Lithium-Lanthanum Titanate Synthesized by Sol-gel Method

Kunshina G.B., Efremov V.V., Ivanenko V.I.

(I.V.Tananaev Institute of Chemistry and Technology of Rare Elements and Mineral Raw Materials of the Kola Science Centre of the Russian Academy of Sciences, Apatity, Murmansk reg., Russia)

ABSTRACT: A new effective method is proposed for synthesis of the $Li_{3x}La_{2/3-x}TiO_3$ oxide solid electrolyte with high lithium ionic conductivity by citrate sol-gel technique. The advantage of the method consists in use of freshly precipitated hydrated titanium hydroxide readily soluble in HNO_3 as a titanium-containing component. The chemical interaction at using of citrate precursor results in a target product for 1 stage owing to a better homogenization of the reaction mixture. This makes it possible to diminish considerably the synthesis temperature (to 1000℃) and duration of preparation of $Li_{3x}La_{2/3-x}TiO_3$ powders. The conditions are determined for production of $Li_{3x}La_{2/3-x}TiO_3$ polycrystalline solid electrolyte with the maximum bulk ionic conductivity of 1.3×10^{-3} S/cm at room temperature. It is shown that the density and a microstructure (grain size) of the samples influence on $Li_{3x}La_{2/3-x}TiO_3$ ion- conductive properties.

KEY WORDS: solid electrolytes; lithium-lanthanum titanate; synthesis; sol-gel method; electrochemical impedance; lithium-ionic conductivity

Silicon-Glycerol-Polysaccharide Systems in Creating Topical Application Drugs

Elena Yu. Larchenko[1], Elena V. Shadrina[1], Marina V. Ryaposova[2], Dmitry M. Kadochnikov[2], Maria N. Isakova[2], Polina B. Zhovtyak[3], Sergey S. Grigoryev[3], Tat'yana G. Khonina[1,4]

(1. Postovsky Institute of Organic Synthesis of Russian Academy of Sciences (Ural Branch), Ekaterinburg 620990, Russia; 2. Ural Scientific Research Veterinary Institute, Ekaterinburg 620142, Russia; 3. Ural State Medical University, Ekaterinburg 620219, Russia; 4. Ural State Agricultural University, Ekaterinburg 620075, Russia)

ABSTRACT: Novel silicon-glycerol-polysaccharide hydrogels on the basis of silicon tetraglycerolate and combined silicon dimethyl- and tetraglycerolates have been synthesized by sol-gel method using polysaccharides (carboxymethyl cellulose, xanthan gum, and hydroxyethyl cellulose) as templates and properties modifiers. The influence of the polysaccharide concentration on the gelation process in comparison with chitosan has been established; their accelerating effect has been shown. A number of agents for medicine and veterinary in liquid and soft dosage forms using silicon glycerolates and polysaccharides has been created.

KEY WORDS: polysaccharides; silicon–glycerol precursor; sol–gel processing; drugs

Controllable Modification of Glassy Composites with Ion-Exchange Technique

A. A. Lipovskii[1], A. V. Redkov[2], A.A. Rtischeva[1], V. V. Rusan[3], D. K. Tagantsev[1], V. V. Zhurikhina[1]

(1. Peter the Great St. Petersburg Polytechnic University, Russia; 2. Institute of Problems of Mechanical Engineering RAS, Russia; 3. Research and Technological Institute of Optical Materials Science Russia)

List of Articles for Publication

Structure and Corrosion Resistance of Ti-0.16 Pd Alloy after Equal Channel Angular Pressing

Lotkov A.I.[1], Kopylov V.I.[2], Latushkina S.Yu.[2], Grishkov V. N.[1], Baturin A.A.[1,2], Korshunov A.V.[3], Abramova P.V.[3], Girsova N.V.[1], Timkin V.N.[1], Zhapova D.Yu.[1]

(1. Institute of Strength Physics and Materials Science SB RAS, Russia; 2. Physical Technical Institute NASB, Belarus; 3. Tomsk Polytechnic University, Russia)

ABSTRACT: The paper presents research data demonstrating the influence of equal channel angular pressing (ECAP) at 650 K on the structural-phase state of a Ti–0.16Pd–0.14Fe alloy (wt.%) with initially microcrystalline structure (<d> = 10 μm). The data show that after equal channel angular pressing, the alloy reveals microinhomogeneities in its chemical and phase composition on grain scales. The central regions of grains is structured as a hexagonal α phase with a low content of Fe and Pd. Near the grain boundaries and junctions, the concentration of Fe and Pd is much higher, which provide the presence of orthorhombic α″ martensite and/or two-phase α+α″ state in these regions. At the grain boundaries and junctions, the material reveals bcc β precipitates (0.6–3.5 μm) containing Fe and Pd in amount of 6–14 and 0.9–1.9 wt.%, respectively. When exposed to equal channel angular pressing, the alloy assumes submicrocrystalline structure (<d> = 0.28 μm) and inherits features associated with grain boundary Fe and Pd segregations. The electrochemical behavior of alloy samples with initial microcrystalline structure and submicrocrystalline structure after ECAP are studied. The results showed that samples after ECAP exhibited higher corrosion resistance in the medical physiological solution 0,9% NaCl than samples with microcrystalline structure.

Structure and Properties of Self-expanding Intravascular NiTi Stents Doped with Si Ions

Lotkov A.I.[1,2], Kashin O.A.[1], Kudryashov A.N.[3], Krukovskii K.V.[1]

(1. Institute of Strength Physics and Materials Science SB RAS, 2/4 Akademicheskii Ave., Tomsk, Russia; 2. National Research Tomsk State University, 36 Lenin Ave., Tomsk 634050, Russia; 3. Angioline Interventional Device Ltd., 18, Inzhenernaya str., Novosibirsk 630090, Russia)

Influence of the Jahn-Teller Effect on the Structure of Ferroelectrics, Magnets and Multiferroics

Makarov Valery Nikolayevich, Kanygina Ol'ga Nikolayevna

(Orenburg State University, Russian Federation)

Bioactive Materials for Restoration Surgery of Bone Tissue

Medkov M.A., Grishchenko D.N.

(Institute of Chemistry, Far East Branch, Russian Academy of Sciences, Russia)

Directional Evolution of the Structure and Properties of HTSC Tapes under the Influence of Impacts Impulses

Mikhailov B.P.[1], Mikhailova A.B.[1], Borovitskaya I.V.[1], Burkhanov G.S.[1], V.Ya.Nikulin[2], P.V.Silin[2]

(1. A.A.Baikov Institute of Metallurgy and Materials Sciences RAS, Moscow, Russia; 2. Lebedev Physical Institute, RAS, Moscow, Russia)

ABSTRACT: The paper presents the results of studying the effect of plasma and mechanical shocks of various intensities on the structure and superconducting parameters (Tc, Jc, frozen magnetic field, VAC and Ic (B)) of superconducting tapes containing interlayers of Bi-2223 and MgB_2. The tapes are prepared according to the known PIT "powder in tube" technology. To apply strikes on the surface of the tapes, the Plasma Focus unit was used, which allows to make a strike at different distances from the plasma anode (20-50 mm) in an argon atmosphere and two specially designed installations: one for applying mechanical shocks at room temperature and second -with heating up to 500°C. The specific energy of the strikes was changed in the range from 0.1 to 100 J/cm^2. The area of the deposition of plasma impacts was 0.6 cm^2, and for mechanical shocks- 0.16 cm^2. The strikes along the surface of the tapes were applied with different steps from 0.4 up to 1.0 mm. The zone of overlapping of the impacts was 0.2 mm. Conducted experiments showed the possibility of increasing the critical current (up to 50%-60%) for MgB_2 tapes in magnetic fields of 1.5 -3.0 T at 4.2 K. For the samples of Bi-2223 tapes, it was established that due to the impact of mechanical shocks and heat treatment, the structure of superconducting core consolidated, the HTSC layers homogenized and the superconducting parameters increased.

KEY WORDS: HTSC materials; plasma and mechanical shocks; superconducting parameters; ceramic structure; density; impact energy; heat treatment

Annealing Temperature Effect on Protonic Conductivity of Aquivion Like Electrolyte Membranes

Kamila Mugtasimova[1,2], Alexey Melnikov[1,3], Elena Galitskaya[1,4], Alexander Sivak[1], Vitaly Sinitsyn[1,4]

(1. InEnergy Group, Elektrodnaya 12-1, 111524, Moscow, Russia; 2. Institute of General Physics RAS, Vavilova 38, 119991, Moscow, Russia; 3. Lomonosov MSU, Leninskie Gory 1, 119991, Moscow, Russia; 4. Institute of Solid State Physics RAS, Academician Ossipyan 2, 142432, Chernogolovka, Russia)

ABSTRACT: Proton-conducting membranes were obtained by solution casting method from new ionomer Inion (Russian analog of Aquivion). It is shown, that annealing of Inion membranes at temperature range from 160℃ to 170℃ leads to significant increase of specific proton conductivity to values even higher, than those of commercial membrane Nafion NR212. After analysis of experimental data, it was suggested, that the optimum annealing regime corresponds to increased water content of these membranes.

Corrosion Resistance of Biocompatible Layered Composite Materials with Shape Memory Effect in Modeling Media

Nasakina E.O., Baikin A.S., Kaplan M.A., Konushkin S.V., Sergienko K.V., Kovaleva E.D., Kolmakova A.A., Kargin Yu.F., Demin K.Yu., Sevost'yanov M.A., Kolmakov A.G., Simakov S.V.

(Institution of Russian Academy of Sciences A.A. Baikov Institute of Metallurgy and Material Science Russian Academy of Sciences, Russia)

The Technology of Preparation of Decellularized Human Liver Tissue Fragments to Create Cell - and Tissue-Engineered Liver Constructs

Nemets E.A., Kirsanova L.A., Basok Ju.B., Lymareva M.V., Schagidulin M.Ju., Sevastianov V.I.

(V.I. Shumakov Federal Research Center of Transplantology and Artificial Organs of the Ministry of Healthcare of the Russian Federation, Moscow, Russian Federation)

ABSTRACT: One of the problems of creating a bioengineered liver as an alternative to the donor liver transplantation for the treatment of end-stage liver failure, is finding the matrix, capable of performing temporarily the functions of a natural extracellular matrix (ECM) and providing the necessary conditions to maintain the viability and functionality of liver cells. The main disadvantage of resorbable biopolymer matrices is the absence of tissue-specific properties and the impossibility of reproducing the unique structure of the liver ECM. The aim of this work was the development of technology for decellularization of liver tissue fragments while preserving the structural properties of the native liver ECM.

Materials and methods: The decellularization of mechanically grinded human liver fragments was carried out in three changes of buffer solution (pH=7.4) containing 0.1% sodium dodecyl sulfate and increasing concentrations of Triton X100 (1%, 2% and 3%, respectively). While developing the technology, we investigated the effects of duration, rinsing conditions (static, dynamic, rotary system, magnetic stirrer), and methods of liver tissue grinding on the complete removal of cellular elements and detritus and preservation of the liver ECM structure. Slices of decellularized liver tissue samples were stained with hematoxylin and eosin, as well as by Masson's method for the detection of connective-tissue elements.

Results and discussion: Methods of histological analysis showed that the most effective for decellularization and structural preservation of the native human liver ECM is a mode of rinsing the liver fragments for three days at room temperature in static conditions, accompanied by stirring with a magnetic stirrer 2-3 times a day for one hour. A longer time or a greater repetition of stirring is accompanied by increased risk of liver tissue damage. On the basis of the test results the algorithm was obtained for a preliminary study of donor human liver, designed to optimize the process of obtaining decellularized liver tissue fragments.

Conclusion: The algorithm of evaluation of donor human liver for decellularization is suggested and optimal conditions of obtaining decellularized liver tissue fragments are found while preserving the liver ECM structure and removing completely cellular elements and detritus.

KEY WORDS: liver; decellularization; resorbable biopolymer matrix; tissue specific properties; tissue-engineered construct

The Use of Spark Plasma Sintering Method for High-rate Diffusion Welding of Ultrafine-grained α-titanium Alloys with High Strength and Corrosive Resistance

Nokhrin A.V.[1], Chuvil'deev V.N.[1], Boldin M.S.[1], Piskunov A.V.[1], Kozlova N.A.[1], Chegurov M.K.[1], Popov A.A.[1], Lantcev E.A.[1], Kopylov V.I.[1,2], Tabachkova N.Yu.[3]

(1. Lobachevsky State University of Nizhny Novgorod, Russia; 2. Physics and Technology Institute, National Academy of Sciences of Belarus, Belarus; 3. The National University of Science and Technology MISiS, Russia)

Studies Into the Impact of Mechanical Activation Modes on Optimal Solid-phase Sintering Temperature of Nano-and Ultrafine-grained Heavy Tungsten Alloys

Nokhrin A.V.[1], Chuvil'deev V.N.[1], Boldin M.S.[1], Sakharov N.V.[1], Baranov G.V.[2], Popov A.A.[1], Lantcev E.A.[1], Belov V.Yu.[2]

(1. Lobachevsky State University of Nizhny Novgorod, Russia; 2. Russian Federal Nuclear Center All-Russian Research Institute of Experimental Physics, Russia)

Spark Plasma Sintering of Bulk Ultrafine-grained Tungsten Carbide with High Hardness and Fracture Toughness

Nokhrin A.V.[1], Chuvil'deev V.N.[1], Blagoveshchenskiy Yu.V.[2], Boldin M.S.[1], Sakharov N.V.[1], Isaeva N.V.[2], Popov A.A.[1], Lantcev E.A.[1], Belkin O.A.[1], S.P. Stepanov[1]

(1. Lobachevsky State University of Nizhny Novgorod, Russia; 2. A.A. Baykov Institute of Metallurgy and Material Science of RAS, Russia)

Ammonolysis of Magnesiothermic Niobium and Tantalum Powders

V. M. Orlov[1], R. N. Osaulenko[2], D.V. Lobov[2]

(*1. Tananaev Institute of Chemistry and Technology of Rare Elements and Mineral Raw Materials, Kola Scientific Center RAS, Apatity, 184209 Russia; 2. Petrozavodsk State University, Petrozavodsk, Republic of Karelia, 185910 Russia*)

ABSTRACT: The influence of ammonolysis conditions in the 400-900℃ temperature range of mesoporous magnesiothermic tantalum powders with specific surface area of 10, 56 m^2/g and niobium with a specific surface area of 18, 83 and 123 m^2/g on the phase composition and specific surface of the products obtained is studied. The dependence of the products composition on the specific surface of the precursor is established. It is due to the ratio of the amount of surface oxide and metal phase in the powder. As a result, nitrogen-containing compounds with a large specific surface area were obtained.

KEY WORDS: niobium; tantalum; powder; ammonolysis; nitride; specific surface area

Preparation and Properties of Nanostructured Materials Based on Polyphenoxazine and Bimetallic Co-Fe Nanoparticles

Ozkan S.Zh., Karpacheva G.P.

(*A.V. Topchiev Institute of Petrochemical Synthesis of Russian Academy of Sciences, Moscow, Russia*)

ABSTRACT: A simple method for preparation of hybrid magnetic materials consisting of bimetallic Co-Fe nanoparticles and polyphenoxazine (PPOA) is described. The nanocomposites were prepared by IR heating of precursors based on PPOA, cobalt (II) acetate and iron (III) chloride in an inert atmosphere at T = 400–600℃ in a short time (2–10 min). Based on the value of hysteresis loop squareness (k_S = 0.021–0.034), it was concluded that almost 100% of Co-Fe nanoparticles in the nanocomposite are superparamagnetic.

KEY WORDS: polyphenoxazine; conjugated polymer; IR heating; metal-polymer nanocomposite; magnetic material; bimetallic Co-Fe nanoparticles

Synthesis Methods of Hybrid Magnetic Materials Based on Polyphenoxazine and Fe_3O_4 Nanoparticles

Ozkan S.Zh., Karpacheva G.P.

(*A.V. Topchiev Institute of Petrochemical Synthesis of Russian Academy of Sciences, Moscow, Russia*)

ABSTRACT: Hybrid metal-polymer nanocomposite materials based on polyphenoxazine (PPhOA) and Fe_3O_4 nanoparticles were obtained for the first time via two methods: *in situ* oxidative polymerization of phenoxazine (PhOA) in an aqueous solution of isopropyl alcohol with nanoparticles of Fe_3O_4 being present; chemical transformations of PPhOA subjected to IR heating at 400–450℃ in the presence of $FeCl_3 \cdot 6H_2O$ in an inert atmosphere. The chemical structure, phase composition, magnetic and thermal properties of obtained nanocomposites were investigated in relation to the synthesis conditions.

KEY WORDS: polyphenoxazine; conjugated polymers; oxidative polymerization *in situ*; IR heating; metal-polymer nanocomposite; magnetic material; Fe_3O_4 nanoparticles

Comparative Kinetics of Extraction of Molybden-containing Components from the Waste Hydrotreatment Catalyst by Mineral Acids

S.P. Perekhoda[1], E.Yu. Nevskaya[2], O.A. Egorova[2]

(*1. Baikov Institute of Metallurgy and Materials Science, RAS, Moscow; 2. Peoples' Friendship University of Russia. RUDN University, Moscow*)

Feathers of Formation of Vinul Siloxane Functional Nanoiayers on Zinc Surface

Maxim Petrunin, Natalia Gladkikh, Ludmila Maksaeva, Marina Maleeva, Elena Terekhova

(*Frumkin Institute of Physical Chemistry and Electrochemistry RAS, Russian*)

ABSTRACT: The method of quartz crystal microbalance is used to study adsorption of vinyl trimethoxysilane (VS) on the surface of zinc from an aqueous solution. Adsorption isotherms are obtained. Approaches corresponding to the known adsorption isotherms are used for interpretation of adsorption data: Langmuir, BET, Flory–Huggins, Langmuir multicenter, Temkin, and Langmuir–Freundlich. It is shown that silanes are adsorbed on the surface of thermally deposited zinc from aqueous solutions and displace adsorbed water from the surface by occupying more than six adsorption sites on the surface. It is found that monolayer coverage of the zinc surface is reached at a concentration of the VS solution of 1×10^{-4} M. The neighboring adsorbate molecules can interact, forming siloxane dimers and trimers bound to the metal surface by either covalent or hydrogen bonds. Adsorption heats are calculated using different adsorption models. It is shown that VS is chemosorbed on the surface of zinc. An increase in the concentration of the VS solution up to 0.1 M results in formation of polycondensed siloxane oligomers on the surface with polycondensation degree n = 8~12. Oligomer surface fragments are connected with each other by hydrogen bonds and are connected with the surface by Zn–O–Si bridge bonds. The overall thickness of such a layer is 10–12 nm or ten molecular layers.

Action of Random Signal on Oscillatory Circuit with Ferroelectric Negative Capacitance

A.A. Potapov[1,2], A.E. Rassadin[3], A.A. Tronov[3]

(*1. Kotel'nikov Institute of Radio Engineering and Electronics of Russian Academy of Sciences, Russia*; *2. Joint-Lab. of JNU-IREE RAS, JiNan University, China*; *3. Nizhny Novgorod Mathematical Society, Russia*)

On Solution of Fokker-planck-kolmogorov Equation for a Ferroelectric Capacitor with a Negative Capacitance by Means of the Krasovsky Series Expansion Method

A.A. Potapov[1,2], I.V. Rakut[3], A.E. Rassadin[4], A.A. Tronov[4]

(1. Kotel'nikov Institute of Radio Engineering and Electronics of Russian Academy of Sciences, Russia; 2. Joint-Lab. of JNU-IREE RAS, JiNan University, China; 3. Lobachevsky State University of Nizhny Novgorod, Russia; 4. Nizhny Novgorod Mathematical Society, Russia)

On Transient Response of *p-n* Junction on Some Ultrawideband Signals

A.A. Potapov[1,2], A.E. Rassadin[3], A.V. Stepanov[4], A.A. Tronov[3]

(1. Kotel'nikov Institute of Radio Engineering and Electronics of Russian Academy of Sciences, Russia; 2. Joint-Lab. of JNU-IREE RAS, JiNan University, China; 3. Nizhny Novgorod Mathematical Society, Russia; 4. Chuvash State Agricultural Academy, Russia)

Generators of Chaotic Electrical Oscillations on Basis of Ferroelectric Capacitor with a Negative Capacitance

A.A. Potapov[1,2], A.E. Rassadin[3], A.A. Tronov[3]

(1. Kotel'nikov Institute of Radio Engineering and Electronics of Russian Academy of Sciences, Russia; 2. Joint-Lab. of JNU-IREE RAS, Jinan University, China; 3. Nizhny Novgorod Mathematical Society, Russia)

Nonlinear Dynamics of Fractals with Cylindrical Generatrix on Surface of Solid State

A.A. Potapov[1,2], A.E. Rassadin[3], A.V. Stepanov[4], A.A. Tronov[3]

(1. Kotel'nikov Institute of Radio Engineering and Electronics of Russian Academy of Sciences, Russia; 2. Joint-Lab. of JNU-IREE RAS, JiNan University, China; 3. Nizhny Novgorod Mathematical Society, Russia; 4. Chuvash State Agricultural Academy, Russia)

Investigation of Stability of Solitary Charge Wave in Infinite Transmission Line with Ferroelectric Capacitors with a Negative Capacitance

A.A. Potapov[1,2], I.V. Rakut[3,4], A.E. Rassadin[4]

(1. Kotel'nikov Institute of Radio Engineering and Electronics of Russian Academy of Sciences, Russia; 2. Joint-Lab. of JNU-IREE RAS, Jinan University, China; 3. Lobachevsky State University of Nizhny Novgorod, Russia; 4. Nizhny Novgorod Mathematical Society, Russia)

Fractal Radioelement's, Devices and Fractal Systems for Radar and Telecommunications

Potapov A.A.[1,2], Potapov Alexey A.[2], Potapov V.A.[1,2]

(1. Kotel'nikov Institute of Radio Engineering and Electronics of Russian Academy of Sciences, Russia; 2. Joint-Lab. of JNU-IREE RAS, Jinan University, China)

Magnetoresistance and Strength Properties of HTSC Composite Tape Joints Obtained by Soldering

Prosvirnin D.V.[1], Troitskii A.V.[2], Markelov A.V.[3], Demikhov T.E.[4], Antonova L.Kh[2], Mikhailov B.P.[1], Molodyk A.A.[3], Mikhailova G.N.[2]

(*1. Institute of Metallurgy and Material Science, RAS, Moscow, 119991, Russia*; *2. Prokhorov General Physics Institute, RAS, Moscow, 119991, Russia*; *3. SuperOx Company, Moscow, 117246, Russia*; *4. Lebedev Physics Institute, RAS, Moscow, 119991, Russia*)

Modes for Ceramics Based on Aluminum Oxynitride, and Their Influence on the Properties of the Material

Prosvirnin D.V.[1], Kolmakov A.G.[1], Alikhanyan A.S.[2], Samokhin A. V.[1], Antipov V.I.[1], Larionov M. D.[1], Titov D.D.[1]

(*1. A.A. Baikov IMET RAS, Moscow, 119334, Russia*; *2. IGIC RAS, Moscow, 119071, Russia*)

ABSTRACT: A comparative analysis of the physicomechanical characteristics and phase composition of $Al_{23}O_{27}N_5$ ceramics obtained by reactive sintering of a mixture of Al_2O_3 and AlN powders at different temperatures and pressures is carried out. The samples obtained had properties that slightly deviate from the literature data.

KEY WORDS: aluminum oxynitride; structure; strength; hardness; speed of sound; reaction sintering

Investigation of the Surface of Destruction of Aluminum Coatings at Negative Temperatures

Prygaev A., Vyshegorodtseva G., Buklakov A., Morozova V.

(*Federal State Budgetary Educational Institution of Higher Education 《Gubkin Russian State University of Oil and Gas (National Research University)》*)

Use of Functional Properties of Nithinol Alloys for Increasing the Energy Efficiency of Pumping Equipment

Prygaev Alexander K.[1], Dubinov Yuri S.[2], Dubinova Olga B.[3], Fatkhutdinov Ruslan R.[4], Nakonechnaya Ksenia V.[5]

(1. Federal State Budgetary Educational Institution of Higher Education《Gubkin Russian State University of Oil and Gas (National Research University)》, Russian Federation; 2. Federal State Budgetary Educational Institution of Higher Education《Gubkin Russian State University of Oil and Gas (National Research University)》, Russian Federation; 3. Federal State Budgetary Educational Institution of Higher Education《Gubkin Russian State University of Oil and Gas (National Research University)》, Russian Federation; 4. Federal State Budgetary Educational Institution of Higher Education《Gubkin Russian State University of Oil and Gas (National Research University)》, Russian Federation; 5. Federal State Budgetary Educational Institution of Higher Education《Gubkin Russian State University of Oil and Gas (National Research University)》, Russian Federation)

From Nano-to Femtotechnology and Back. About Possibility of Obtaining a Neutron Substance in Laboratory Conditions

Ryazantsev G.B.[1], Beckman I.N.[1], Lavrenchenko G.K.[2], Khaskov M.A.[3], Pokotilovskiy Yu. N.[4]

(1. Lomonosov Moscow State University, Leninskie Gory, Moscow, Russia; 2. LLC《Institute of Low Temperature Energy Technology》, POB188, Odessa, Ukraine; 3. All-Russian Scientific Research Institute of Aviation Materials, Moscow, Russia; 4. Joint Institute for Nuclear Research, Dubna, Russia)

ABSTRACT: Considering the formation of a neutron substance, besides other gravitational neutronization, other mechanisms, such as condensation of ultracold neutrons (UCN) and neutronization due to a critical increase in the atomic number in the periodic table (PS). The stability of the neutron substance is substantiated already at the micro level due to Tamm interaction and not only at the macro level due to the gravitational interaction, as it is now considered in astrophysics. A neutron substance is a very concrete physical reality, urgently demanding its rightful place in the PS and studying not only physical, but also chemical, and possibly even in the near future, engineering and technical properties. We also consider the possibility of a "chemical" interaction of UCN with molecules of substances with an odd number of electrons. It is proposed to extend the PS beyond the limits of classical chemical substances and to cover a much wider range of matter in the universe, based on the forgotten ideas of D.I. Mendeleev. Moreover, begins with neutron and its isotopes (dineutron, tetrautronitrone, etc.) and ends PS the neutron stellar substance.

KEY WORDS: neutron; a neutron substance; inert gases; periodic system of elements; neutronization; zero period

Control of Powders Properties under Conditions of Synthesis and Treatment in DC Thermal Plasma Jet

Samokhin A. V., Alekseev N. V., Kirpichev D. E., Astashov A. G., Fadeev A. A., Sinaiskiy M. A., Tsvetkov Y. V.

(A. A. Baikov Institute of Metallurgy and Materials Sciences, Russian Academy of Sciences Russia, 119334, Moscow, Leninskiy prospect 49)

ABSTRACT: A review of the research results on the synthesis and processing of powders in DC thermal plasma flows is presented. Mechanisms for the formation of metal nanoparticles and their compounds under plasma flow conditions are considered. The results of experimental studies on the effect of the plasma process parameters on the properties of powders are presented. A complex system for the powders analysis has been developed — particle size distribution, average size, phase composition, impurities, etc. The presented results of research and development testify to the wide possibilities of plasma processes and devices for obtaining and processing powders of metals and their inorganic compounds with specified properties. Produced nanopowders are used in R & D to create new materials with special and improved properties.

Calculation of Direct and Inverse Current-voltage Characteristics of Schottky Barrier Height and on the Border of Metal-semiconductor in a Nonlinear Model for Diodes Based on SiC

A.V. Sankin[1], V.I. Altukhov[1], A.S. Sigov[2], S.V. Filipova[1], E. G. Janukjan[1]

(1. North-Caucasian Federal University, 355009, Russia, Stavropol; 2. Moscow Institute of Radio Electronics and Automatics, 119454, Russia, Moscow)

ABSTRACT: For solid solutions SiC heterostructures and diode type Al/SiC-AlN proposed nonlinear surface densities of States of Schottky barrier model with resonant quasi-level Fermi. Calculated height barrier Schottky (BSH) used to calculate the direct and inverse current-voltage characteristics of diode type M/n–$(SiC)_{1-x}$ $(AlN)_x$.

KEY WORDS: barrier height schottky; diodes based on SiC; direct and inverse current-voltage characteristics

Formation of Composite Materials Depending on the Geometrical Parameters

Seregin A.V., Nasakina E.O., Sudarchikova M.A., Sprygin G.S., Khimyuk Ya.Ya., Demin K.Yu., Kaplan M.A., Konushkin S.V., Fedyuk I.M., Yakubov A.D., Sevost'yanov M.A., Kolmakov A.G.

(Institution of Russian Academy of Sciences A.A. Baikov Institute of Metallurgy and Material Science Russian Academy of Sciences, Moscow, Russia)

Technology of Tissue Engineering and Regenerative Medicine in the Treatment of Damaged Articular Cartilage

V.I. Sevastianov[1], Yu.B. Basok[1], A.M. Grigoryev[1], L.A. Kirsanova[1], V.N. Vasilets[1], S.V. Gautier[1,2]

(1. Shumakov Research Center of Transplantology and Artificial Organs, Moscow, Russia Federation; 2. I.M. Sechenov First Moscow State Medical University, Department of Transplantology and Artificial Ograns, Moscow, Russia Federation)

Simulation of Stress-strain State of Pipeline Parts under Plastic Deformation and Fracture

Shabalov I.P.[1], Velikodnev V.Y.[1], Murzakhanov G.Kh.[2], Kalenskii V.S.[1], Nastich S.U.[1], Barsukov A.A.[2], Arsenkin A.M.

(1. Pipe Innovation Technologies, LLC, Moscow, Russia; 2. Moscow Research Center of Structural Materials of JSC "MOSGAZ", Moscow, Russia)

Promising Ceramic Materials Based on Elementoxane Precursors

Shcherbakova G.I.[1], Varfolomeev M.S.[2], Krivtsova N.S.[1], Storozhenko P.A.[1], Moiseev V.S.[2], Yurkov G.Yu.[3], Ashmarin A.A.[3]

(*1. SSC RF State Research Institute for Chemistry and Technology of Organoelement Compounds*;
2. Moscow Aviation Institute (*National research university*), *Russia*;
3. Baikov Institute of Metallurgy and Material Science RAS)

Research of Thermophysical Properties of Organoplastics Based on Polyarylate Sulfone Block Copolymer

Shustov G. B.[1], Burya A. I.[2], Grashchenkova M. A.[2], Shetov R. A.[1]

(*1. Kh.M.Berbekov Kabardino-Balkarian State University, Russia*; *2. Dneprodzerzhinsk State Technical University, Ukraine*)

ABSTRACT: Organoplastics based on polyarylate sulfone block copolymer, reinforced with organic fiber terlon, have been developed. The influence of the percentage content of fibrous filler on the specific heat, thermal conductivity and thermal diffusivity of organoplastics is shown. The temperature dependences of these thermophysical characteristics on the basis of which the analysis of structural molecular transformations in the materials under study was made.

KEY WORDS: block copolymer; organoplastics; specific heat; thermal conductivity; thermal diffusivity

Low-temperature Extraction-pyrolytic Synthesis of Nano-size Composites Fased on Metal Oxides

N.I. Steblevskaya, M.A. Medkov, M.V. Belobeletskaya

(Institute of Chemistry, Far-East Branch, Russian Academy of Sciences, Vladivostok, Russia)

Electrochemical Forming of Electroconductive Poly-Fe (III)-Aminophenylporphyrin Films in Various Solvents

Tesakova M.V.[1], Kuzmin S.M.[1,2], Chulovskaya S.A.[1], Semeikin A.S.[3], Parfenyuk V.I.[1,3,4]

(1. G.A. Krestov Institute of Solution Chemistry of RAS, Russia; 2. Ivanovo State Power Engineering University, Russia; 3. Ivanovo State University of Chemistry and Technology, Russia; 4. Kostroma State University, Russia)

Role of Graphene Sheets in Formation of Metal Oxides Based Hybrid Nanostructures

Trusova E.A.[1], Kirichenko A.N.[2], Kotsareva K.V.[1], Polyakov S.N.[2], Abramchuk S.S.[3]

(1. Institution of the Russian Academy of Sciences A.A. Baikov Institute of Metallurgy and Materials Science of RAS, Moscow, Russia; 2. Technological Institute for Superhard and Novel Carbon Materials, Troitsk, Moscow, Russia; 3. Lomonosov Moscow State University, Moscow, Russia)

ABSTRACT: The combined method including sol-gel and sonochemical techniques is proposed for the synthesis of the hybrid nanostructures based on graphene and metal oxides. It was found that during the synthesis in N,N-dimethyloctylamine-aqua mixture, graphene at first is a structuring agent of the reaction mixture, and then its sheets are textured component of a hybrid nanostructure.

Effect of the Power Pulses of Deuterium Plasma on the Structure and Mechanical Properties of V-Ga Alloys

A.B. Tsepelev[1,2], V.F. Shamray[1], V.P. Sirotinkin[1], N.A. Vinogradova[1]

(*1. A.A. Baikov Institute of Metallurgy and Materials Science of the Russian Academy of Sciences (IMET RAS); 2. National Research Nuclear University MEPhI*)

ABSTRACT: Effects of the power nanosecond pulse flows of deuterium ions and dense deuterium plasma (10^8 W/cm^2) produced by Plasma Focus device on the structure and mechanical properties of V-base alloys have been investigated. It is shown that V-Ga-Cr alloys have a better resistance to radiation embrittlement under radiation-thermal impact compared to V-Ti-Cr ones. X-ray diffraction analysis data indicates that the solid solutions on the base of V are stable under a plasma irradiation.

Highly Effective Activated Carbon from Wood for Supercapacitors: Synthesis and Research

Vervikishko D.E., Kochanova S.A., Kiselyova E.A., Shkolnikov E.I.

(*Joint Institute for High Temperatures RAS, Moscow, Russia*)

Recent Advances in Laser-Pulse Melting of Graphite at High Pressure

P.Vervikishko, M. Sheindlin

(*Joint Institute for High Temperatures of the Russian Academy of Sciences*)

The Influence of Aluminium Addition on the Structure and Magnetic Properties of the Pseudobinary $Sm_2(Fe_{1-x}Al_x)_{17}$ Alloys and Their Hydrides

Veselova S.V.[1], Verbetsky V.N. [1], Savchenko A.G.[2], Shchetinin I.V. [2]

(1. Chemistry Department, Lomonosov Moscow State University, Moscow, Russia; 2. Department of Physical Materials Science, National University of Science and Technology 《MISiS》, Moscow, Russia)

ABSTRACT: Results of XRD analysis and magnetic measurements performed on $Sm_2(Fe_{1-x}Al_x)_{17}$ (x=0.1–0.4) samples and their hydrides are reported. X-ray powder diffraction studies using the Rietveld method have shown that aluminium alloys crystallize in the Th_2Zn_{17} type rhombohedral structure and synthesized hydrides preserved the same type structure as their parent compounds. It was established that the replacement of Al for Fe in Sm_2Fe_{17} causes a monotonic increase in the structural parameters for $Sm_2(Fe_{1-x}Al_x)_{17}$ compounds. The specific saturation magnetization (σ_s), specific remanent magnetization (σ_r) and specific coercive force by magnetization (jH_c) of all samples carried out at room temperature (H=9 T) decreases by Al substitution. As well as the magnetic properties of hydrides with analogous tendence has been studied. The $Sm_2(Fe_{1-x}Al_x)_{17}H_y$ compounds were prepared by heat-treating powders of corresponding two-phases initial alloys in H_2 gas. The introduction of hydrogen atoms leads to an increase in the lattice constants and unit cell volume, but the increase becomes less noticeably with raising Al concentration.

KEY WORDS: rare earth compounds; hydrides based on Sm-Fe-Al alloys; crystal structure; X-ray phase analysis; electron probe microanalysis; magnetic measurements

ВЛИЯНИЕ АЛЮМИНИЯ НА СТРУКТУРУ И МАГНИТНЫЕ СВОЙСТВА ПСЕВДОБИНАРНЫХ СПЛАВОВ $Sm_2(Fe_{1-x}Al_x)_{17}$ И ИХ ГИДРИДОВ

Веселова С.В.[1], Вербецкий В.Н.[1], Савченко А.Г.[2], Щетинин И.В.[2]

(1. Химический факультет, Московский государственный университет имени М.В. Ломоносова, Москва, Россия; 2. Кафедра физического материаловедения, Национальный исследовательский технологический университет 《МИСиС》, Москва, Россия)

РЕЗЮМЕ: Сообщается о результатах РФА и магнитных измерений, проведенных на $Sm_2(Fe_{1-x}Al_x)_{17}$ (x=0.1–0.4) образцах и их гидридах. Рентгенографические исследования порошковых образцов с применением метода Ритвельда показали, алюминиевые сплавы кристаллизуются в ромбоэдрическом структурном типе Th_2Zn_{17}, синтезированные гидриды сохранили тот же тип структуры, что и родственные соединения. Установлено, замещение железа алюминием в Sm_2Fe_{17} приводит к монотонному увеличению структурных параметров $Sm_2(Fe_{1-x}Al_x)_{17}$. Удельная намагниченность насыщения, остаточная удельная намагниченность насыщения и остаточная коэрцитивная сила по намагниченности всех образцов, измеренные при комнатой температуре в магнитных полях до 9 Тл, убывают по мере замещения алюминием. Также были проведены измерения магнитных свойств гидридов с аналогично выявленной тенденцией. Соединения $Sm_2(Fe_{1-x}Al_x)_{17}H_y$ получены термообработкой порошков соответствующих исходных двухфазных образцов в атмосфере водорода. Внедрение атомов водорода является причиной увеличения периодов и объема элементарной ячейки, что становится менее заметным с ростом концентрации алюминия.

КЛЮЧЕВЫЕ СЛОВА: редкоземельные соединения; гидриды на основе сплавов Sm-Fe-Al; кристаллическая структура; РФА; электронно-зондовый микроанализ; магнитные измерения

Elemental Analysis of Ceramic Materials of Medical Applying

Volchenkova V.A., Kazenas E.K., Andreeva N.A., Ovchinnikova O.A., Paunov A.K., Penkina T.N., Rodionova S.K., Fomina A.A., Podzorova L.I., Ilyicheva A.A.

(Baikov Institute of Metallurgy and Materials Science of the Russian Academy of Sciences (IMET RAS), Russia)

Potential Output from Technological Deadlock in Creation of New Generation Technique

Georg Volkov

(Moscow Polytechnical University, Russia)

ABSTRACT: For the development and manufacture of competitive engineering products offered theoretical basis and technological principles of creation the engineering materials with applicational properties many times higher than current level. The theory realized on a model system of carbon-carbon . Made bulk carbon nanomaterial with unique properties that allowed to create engineering products with the specifications above world level. These results can be used to create of bulk nanomaterials different chemical composition with no less unique properties.
KEY WORDS: bulk nanomaterials; single step technology; technical potential; carbon-carbon system

Microstructure Investigation of Low Carbon Low Alloy Steel Suitable for Application in Arctic Constructions

Vorkachev K.G., Kantor M.M., Solntsev K.A.

(A.A. Baikov Institute of Metallurgy and Materials Science of RAS, Russia)

Investigation of Wear Resistance of Aluminum and Zinc Coatings at Low Temperatures

Vyshegorodtseva G., Buklakov A., Morozova V., Vyshegorodtseva I.

(Federal State Budgetary Educational Institution of Higher Education 《Gubkin Russian State University of Oil and Gas (National Research University)》)

The New Nanocomposite Carbon Material for High-field Electron Sources

Yafarov R.K., Shanygin V.Ya., Nefedov D.V.

(Kotel'nikov Institute of Radio-engineering and Electronics of RAS (Saratov Branch), Zelenaya str. 38, Saratov, 410019, Russia)

ABSTRACT: Requirements and problems are formulated for the creation of field emission cathode materials for high-current emission electronics. Based on the analysis of literature data, it has been shown that the development of new nanostructured diamond-graphite materials is necessary to create autocathodes with a current density of up to 100 A/cm^2 and higher. These materials should have a surface density of diamond nanocrystallites in a graphite matrix of not less than 10^6 - 10^8 cm^{-2}. With the use of a nonequilibrium low-pressure microwave plasma, regions of regimes providing for separate production of carbon film structures of a given allotropic modification (diamond, graphite, graphene-like) and nanocrystalline structures containing a diamond and a graphite phase in different volume ratios have been determined. For the first time, the effect of self-organization of diamond nanocrystallites in graphite and polymer-like carbon films was observed upon precipitation from low-pressure ethanol vapor using a nonequilibrium highly ionized microwave plasma. A technology has been developed for obtaining nanocomposite diamond-graphite coatings with adjustable thresholds for field emission in the range from 5-7 to 20 V/μm and density of field emission currents above 100 A/cm^2.

KEY WORDS: microwave plasma; auto emission; diamond nanocrystallites; self-organization

Metal Nanowires-New Type of Nanomaterial: Fabrication by Matrix Synthesis Technique and Investigation of Structure and Properties

D.L. Zagorskiy[1], S.A. Bedin[1,2], I.M.Doludenko[3], G.G.Bondarenko[3], K.V. Frolov[1], V.V.Artemov[1], M.A. Chuev[4], A.A.Lomov[4]

(*1. Centre of Crystallography and Photonics RAS, Moscow, Leninskii pr. 59;*
2. Moscow Pedagogical State University, Moscow, Malaya Pirogovskayast 1/1;
3. National Research University Higher School of Economics, Moscow, Myasnitskaya Ulitsa 20;
4. Institute of Physics and Technology RAS, Moscow, Nahimovskiyprosp., 36/1)

ABSTRACT: Ensembles of Nanowires (NW) of iron group metals-pure metals (Fe, Ni and Co) and their alloys (Fe-Ni, Fe-Co) were obtained using matrix synthesis technique based on polymer track matrixes. Compositions of electrolytes were chosen – the salt of one corresponding metal (in the first case) and two salts (for second case). The galvanic process was investigated and it was found that it consists of different stages. Deposition of metal inside the pores has non-linear character due to diffusion limitation. The specific features of the next part (formation and growing of the "caps") was also studied. Electron microscopy, X-rays analysis, Mössbauer spectroscopy and magnetic hysteresis were applied to investigate the dependence of structure and magnetic properties of the NW on electrodeposition conditions. It was found that the composition of two-component NWs differs from the composition of electrolyte and different at different parts of NW. Mössbauer spectroscopy gave possibility to estimate hyperfine parameters for Fe-Co NWs. For Fe-Ni NWs it was supposed that the spectra could be presented as superposition of at least three magnetic sextets with hyperfine parameters B_{hf} 27-33 T. It was shown that Fe-Co samples have "hard magnetic" properties, while Fe-Ni samples have "soft magnetic" parameters. The dependence of these parameters on the synthesis was demonstrated.

Membrane Palladium-Based Alloys for High Purity Hydrogen Production

G.S. Burkhanov, N.R. Roshan, E.M. Chistov, T.V. Chistova, S.V. Gorbunov, A.D. Zakharov

(*Baikov Institute of Metallurgy and Materials Science, Russian Academy of Sciences*)

ABSTRACT: In this article advantages of membrane technology for high purity hydrogen production are described. Applying author's background and experience in development of membranes and membrane units with high hydrogen permeability and low Palladium consumption, 20 and 10 micrometer thickness thin foil membranes from effective PdInRu, PdRu, PdCu tailored composition alloys were developed for operation in field gas medias and conditions. Key properties of the membranes were investigated. Technique for membrane sealing with construction elements was developed. Pilot membrane units were manufactured and tested and their hydrogen production efficiency was determined.

Polarization of Glass with Positive and Negative Charge Carriers

V. V. Zhurikhina[1], M. I. Petrov[2], A. A. Rtischeva[1], A. A. Lipovskii[1]

(1. Peter the Great St. Petersburg Polytechnic University, Russia; 2. ITMO University, St. Petersburg, Russia)

Laser Melting and Thermal Treatment of Co-Cr Based Alloys

Drápala J., Losertová M., Konečná K., Kostiuková G.

(Vysoká škola báňská – Technical University of Ostrava, Faculty of Metallurgy and Materials Engineering, Av. 17. Listopadu 15, 70833 Ostrava – Poruba, Czech Republic, EU)

ABSTRACT: Co-Cr based alloys usually produced by conventional casting or newly by Selective Laser Melting (SLM) have been widely used in dentistry as dental prostheses. This work is focused on the size and chemical composition of initial powder particles of a commercial Co-Cr-Mo-W alloy, as well as on the microstructure after SLM and heat treatment. Material used in this study was in a fine powder form (EOS Cobalt Chrome SP2 type). Its composition corresponds to type 4 CoCr dental materials in EN ISO 22674:2006 standard. It also fulfils the chemical and thermal requirements of EN ISO 9693 for CoCr PFM (porcelain fused metal). In order to solve the problem of porosity and shape stability of prostheses after SLM and high temperature annealing, microstructure and phase evaluation using optical and scanning electron microscopies completed by EDX analysis were performed.

KEY WORDS: Co-Cr-Mo-W alloy; selective laser melting; powder; heat treatment; microstructure; porosity; chemical composition

Thermal Effects in Composites Hydroxyapatite-polysaccharide

Svetlana A. Gerk, Olga A. Golovanova

(F.M. Dostoevsky Omsk State University, Russia)

Crystallization of Carbonated-hydroxyapatite and Si-hydroxyapatite on Titanium Implants

Olga A. Golovanova

(F.M. Dostoevsky Omsk State University, Russia)

ABSTRACT: Carbonated-hydroxyapatite (CHA) and Si-hydroxyapatite (Si-HA) precipitation have been synthesized from the model bioliquid solutions (synovial fluid and SBF). It is found that all the samples synthesized from the model solutions are single-phase and represent hydroxyapatite. The crystallization of the modified hydroxyapatite on alloys of different composition, roughness and subjected to different treatment techniques was investigated. Irradiation of the titanium substrates with the deposited biomimetic coating can facilitate further growth of the crystal and regeneration of the surface.

Nanostructuring of the Bismuth Single Crystal Surface (111) under the Action of Atomic Hydrogen

O.I. Markov, Yu.V. Khripunov

ABSTRACT: We studied the morphology of the cleavage surface of a bismuth single crystal (111) by the method of atomic force microscopy after its treatment in an atomic hydrogen atmosphere. We determined the nanostructuring of the surface as a result of formatting micro- and nanocrystalline formations in the form of triangular pyramids. We researched the equilibrium form of nano-formations, as well as their statistical parameters.
KEY WORDS: atomic force microscope; surface; bismuth

Creation of Competitive Composite Materials in the Presence of Interphase Layer on the Border of the Inclusion and the Matrix

Pavlov S. P., Bodyagina K. S.

(Russia, Yuri Gagarin State Technical University of Saratov)

ABSTRACT: In this paper, composites made of periodically repeating the micro structures are investigated. The study aims at identifying the optimal spatial distribution of constituents within a composite material so as to obtain the material desired/improved functional properties. To find the relationship between micro - and macro-structural properties of the composite material, the method of homogenization used. The problem of finding optimal microstructures for materials with maximum rigidity, in the form of a bulk modulus or shear modulus for the basic cell of a composite, for a new class of problems with a microstructure consisting of a set of different preset materials surrounded by an interphase layer was first investigated.

KEY WORDS: composites; homogenization; topology optimization; extreme elastic properties; finite element method

Study on the Microstructure and Properties of $Ag\text{-}SnO_2$ and $Ag\text{-}CuO\text{-}La_2O_3$ Electrical Contact Materials Prepared by Powder Metallurgy

Xie Ming, Wang Song, Zhang Jiming, Li Aikun, Hu Jieqiong, Wang Saibei, Yang Youcai, Chen Yongtai, Liu Manmen

(Kunming Institute of Precious Metals, Sino-platinum Metal Co., Ltd., State Key Laboratory of Advanced Technologies for Comprehensive Utilization of Platinum Metals, Kunming 650106, China)

ABSTRACT: Using chemical coprecipitation, cold isostatic pressing, vacuum sintering and hot isostatic pressing technology integration, the $Ag\text{-}SnO_2$ and $Ag\text{-}CuO\text{-}La_2O_3$ electrical contact materials were fabricated, and the microstructures and properties were investigated. The results show that the oxides are evenly distributed in the contact materials, and the density, resistivity and micro-hardness of $Ag\text{-}CuO\text{-}La_2O_3$ materials are 9.77 $g \cdot cm^{-3}$, 2.36 $\mu\Omega \cdot cm$ and HV113 respectively. Compared with convenient $Ag\text{-}SnO_2$ electrical contact materials prepared by the same process, the $Ag\text{-}CuO\text{-}La_2O_3$ material has superior room temperature processing performance, better ability of anti-arc erosion and longer service life. The $Ag\text{-}CuO\text{-}La_2O_3$ material may be substitute $Ag\text{-}SnO_2$ as a new electrical contact material.

KEY WORDS: electrical contact material; powder metallurgy; $Ag\text{-}SnO_2$; $Ag\text{-}CuO\text{-}La_2O_3$; microstructure; property

Hydrogen Effects on Different Properties of Biocompatible Metallic Materials

Losertová M.

(Regional Materials Science and Technology Centre, Faculty of Metallurgy and Materials Engineering, VŠB-Technical University of Ostrava, Av. 17. Listopadu 15, 708 33 Ostrava-Poruba, Czech Republic)

ABSTRACT: The comparative study of hydrogen effect was performed for different metallic materials in non-hydrogenated and hydrogenated stages: AISI 316L stainless steel, Ti_6Al_4V and $Ti_{26}Nb$ alloys. First, the effect of heat treatment under hydrogen gas on microstructure and room temperature tensile properties was investigated for 316L austenitic stainless steel and Ti_6Al_4V alloy. The hydrogenated 316L steel with the hydrogen content of 17.5 wt. ppm showed the higher yield strength and ultimate tensile stress, unlike Ti_6Al_4V, which the mechanical properties lowered due to the hydrogen amount as high as 2235 wt. ppm.

Second, the positive effect of hydrogen on microstructure and hot deformation behavior at 700 and 750℃ was investigated after thermal hydrogen treatment for Ti_6Al_4V and $Ti_{26}Nb$ alloys. Comparing the results obtained for the non-hydrogenated and hydrogenated specimens of both alloys, it was found that the hydrogen content as high as 1325 wt. ppm has an obvious benefit effect on high temperature deformation behavior in the Ti_6Al_4V alloy by stabilizing beta phase and lowering the thermal deformation resistance. In the case of $Ti_{26}Nb$ alloy the hydrogen content of 2572 wt. ppm suppressed stress instabilities during hot compression but slightly increased the thermal deformation resistance.

The microstructure study was performed before and after the tensile tests as well as before and after isothermal compression tests on the specimens in hydrogenated and non-hydrogenated conditions. The amount of hydrogen in the specimens was measured by means of an analyzer LECO RH600.

KEY WORDS: hydrogen effect; hydrogen embrittlement; hydrogen induced plasticity; biocompatible alloys

Research of Reacting System Self-Purification in the Process of Self-Propagating High-Temperature Synthesis (SHS)

A.A. Potekhin, D.A.Gorkaev, A.Yu. Postmikov, A.I. Tarasova, A.Ya. Malyshev

(Russian Federal Nuclear Center — All-Russian Research Institute of Experimental Physics (RFNC-VNIIEF), 37, Mira Ave., 607188, Sarov, Nizhny Novgorod region, Russia)

ABSTRACT: It was studied the distribution of non-metallic admixtures, i.e. oxygen and nitrogen, in titanium hydride samples obtained using the SHS method in different facilities. It was shown that the least content of the controlled admixture elements was in the central, i.e. the most heated part of the sample.

Besides, it was researched the process of titanium hydride self-purification during SHS. The hypothetical mechanism of the aforementioned process is proposed and the assumption is formulated with respect to the influence of the specific surface size on the degree of titanium hydride self-purification.

KEY WORDS: titanium hydride; titanium powder; self-purification; oxygen; nitrogen; gas content; admixture distribution

The Use of Analysis Spectral Methods to Assess Powder Compositions for Self-Propagating High-Temperature Synthesis

A. Yu. Postnikov, V.V. Mokrushin, A.A. Potekhin, I.A. Tsareva, O.Yu. Yunchina, M.V. Tsarev, D.V. Chulkov, P.G. Berezhko

(Russian Federal Nuclear Center — All-Russian Research Institute of Experimental Physics (RFNC-VNIIEF), 37, Mira Ave., 607188, Sarov, Nizhny Novgorod Region, Russia)

ABSTRACT: Spectral methods were used to analyze the change of variation coefficients V_{mic} and V_{mac} depending on the time of Ti+Al batch mixing and the sizes of spaces excited by laser and electron beams. It is shown that the investigation of mix homogeneity and mixing quality by the methods of spectral analysis with the high level of localization is impossible without comparison between the effect of analytic sample sizes and the size of space of analytic signal excitation.

The Ti+Al SHS system was used to study the change of a phase composition of combustion products depending on original batch mixing time. It is shown that when increasing mixing time from 5 to 20 minutes, the fraction of nonstoichiometric titanium aluminide in the process of synthesis decreases significantly, and when the blend with a 20-minute mixing period burns, the formation of predominately stoichiometric titanium monoaluminide is observed.

KEY WORDS: spectral methods; homogeneity; SHS system; coefficient of variation

Study of Phase Diagram and Thermodynamic Parameter of Au-Sn-Pt System

Xie Ming, Hu Jieqiong, Zhang Jiming, Wang Saibei, Li Aikun, Liu Manmen, Yang Youcai, Chen Yongtai, Fang Jiheng

(*Kunming Institute of Precious Metals, Sino-platinum Metal Co., Ltd., State Key Laboratory of Advanced Technologies for Comprehensive Utilization of Platinum Metals, Kunming 650106, China*)

ABSTRACT: Rare and precious metals materials with unique physical and chemical properties, are widely used in aviation, aerospace, navigation, electronic, electric, petroleum chemical industry, glass fiber, waste gas purification, metallurgical and other industries, thus known as "the vitamin of modern industry". In this paper choose the Au-Sn-Pt ternary alloy as the research object, adopts Pandat software, systematically studied the composition of binary and ternary alloy phase diagram in the Au-Sn-Pt system, and the related thermodynamic parameters, including the phase diagram analysis, phase structure and formation enthalpy of alloy, the cohesive energy and the Gibbs free energy, at the same time, through the analysis and test methods such as XRD, SEM has carried on the experimental verification of phase composition and the phase structure of alloy, predicted the formation mechanism of the stable phase, provides the theoretical and technical basis for the system application in catalytic materials and brazing materials.

Application of Metal Hydrides as Pore-Forming Agents for Obtaining Metal Foams

N.V. Anfilov, A.A. Kuznetsov, P.G. Berezhko, A.I. Tarasova, I.A. Tsareva, V.V. Mokrushin, M.V. Tsarev, I.L. Malkov

(*Russian Federal Nuclear Center — All-Russian Research Institute of Experimental Physics (RFNC-VNIIEF), 37, Mira Ave., 607188, Sarov, Nizhny Novgorod Region, Russia*)

ABSTRACT: One of the well-known ways of metal foam production consists of adding powder which releases gaseous decomposition product at high temperatures into a melted metal. When producing aluminum foams, titanium hydride powder is most frequently used, while its main disadvantage is its insufficient thermal stability. This results in a premature gas loss in the process of the powder introduction into a melted metal matrix. The paper discloses approaches to suppression of the undesirable premature titanium hydride decomposition by means of its preliminary thermal treatment, as well as the external pressure elevation over that of the system being processed during the foam formation. A method of using the thermally treated powder is proposed, in order to obtain aluminum foam samples having a required geometry and sufficiently uniform porous structure.

KEY WORDS: metal foam; aluminum melt; pore-forming agent; pre-oxidized titanium hydride powder; thermal stability; shielding coating; external pressure; porous structure

Application of Acoustic Emission Method to Study Metallic Titanium Hydrogenation Process

A.A. Kuznetsov, P.G. Berezhko, S.M. Kunavin, E.V. Zhilkin, M.V. Tsarev, V.V. Yaroshenko, V.V. Mokrushin, O.Yu. Yunchina, S.A. Mityashin

(Russian Federal Nuclear Center — All-Russian Research Institute of Experimental Physics (RFNC-VNIIEF), 37, Mira Ave., 607188, Sarov, Nizhny Novgorod Region, Russia)

ABSTRACT: The changes occurring in metallic titanium specimens during hydrogenation, that are accompanying by high amplitude acoustic emission signals, have been studied. It was determined that the most probable reason for generation of these signals is crack formation in hydrogenated specimens as a result of internal stresses caused by structural changes in material. Also it was determined that typical sizes of the cracks are comparable with particle sizes of hydrogenated specimens, and the atomic ratio [H]/[Ti] in a solid phase, when crack formation is the most intensive, is generally lower for titanium sponge than for powder made from this sponge, which is explained by a larger specific surface of powder comparatively that of sponge.

KEY WORDS: acoustic emission; hydrogenation; titanium sponge; titanium powder; cracking

Device for Reversible Hydrogen Isotope Storage with Aluminum Oxide Ceramic Case

I.P. Maximkin, A.A. Yukhimchuk, V.V. Baluev, I.L. Malkov, R.K. Musyaev, D.T. Sitdikov, A.V. Buchirin, V.V. Tikhonov

(Russian Federal Nuclear Center — All-Russian Research Institute of Experimental Physics (RFNC-VNIIEF), 37, Mira Ave., 607188, Sarov, Nizhny Novgorod Region, Russia)

ABSTRACT: The work describes the results of testing a trap with the external load-bearing case made of Al_2O_3 ceramics of the brand F99.7 and intended for reversible storage of hydrogen isotopes. When working with the trap, an inductive heater was used. Titanium (11.8g) was the trap's sorption material. The trap's sorption capacity was 4830 cm^3 of protium. As a result of testing it was found that the duration of heating the internal ampoule of the trap containing titanium up to 700℃ was 5 minutes, while the time of its cooling down to the room temperature is 25 minutes. The amount of gas released from the trap at the temperature of 700℃ and protium equilibrium pressure of 740 mBars was 75.6 % of the sorption capacity, while at the temperature of 800℃ and equilibrium pressure of 900 mBars it was 91.9 %.

KEY WORDS: hydrogen pressure; hydrogen isotope storage; titanium hydride; Al_2O_3 ceramics